Ecolinguistics, Social Justice and Sustainability

Bloomsbury Advances in Ecolinguistics

Bloomsbury Advances in Ecolinguistics emerges at a time when businesses, universities, national governments and many other organisations are declaring an ecological emergency. With climate change and biodiversity loss diminishing the ability of the Earth to support life, business leaders, politicians and academics are asking how their work can contribute to efforts to preserve the ecosystems that life depends on.

This book series explores the role that linguistics can play in addressing the great challenges faced by humanity and countless other species. Although significant advances have been made in addressing social issues such as racism, sexism and social justice, linguistics has typically focused on oppression in human communities and overlooked other species and the wider ecosystems that support life. This is despite the disproportionate impact of ecological destruction on oppressed groups. In contrast, this book series treats language as an intrinsic part of both human societies and wider ecosystems. It explores the role that different areas of linguistic enquiry, such as discourse analysis, corpus linguistics, language diversity and cognitive linguistics can play at a time of ecological emergency.

The titles explore themes such as the stories that underpin unequal and unsustainable industrial societies; language contact and how linguistic imperialism threatens the ecological wisdom embedded in endangered languages; the use of linguistic analysis in ecocriticism, ecopsychology and other ecological humanities and social sciences; and emerging theoretical frameworks such as Harmonious Discourse Analysis. The titles also look to cultures around the world for inspirational forms of language that can lead to new stories to live by. In this way, the series contributes to linguistic theory by placing language fully in its social and ecological context, and to practical action by describing the role that linguistics can play in addressing ecological issues.

Titles published in the series:

Corpus-Assisted Ecolinguistics, Robert Poole
Ecolinguistics and Environment in Education, Emile Bellewes
Ecological Communication and Ecoliteracy, edited by Maria Bortoluzzi and Elisabetta Zurru
Exploring Ecolinguistics, Douglas Mark Ponton
Language and Ecology in Southern and Eastern Arabia, edited by Janet C.E. Watson, Jon C. Lovett and Roberta Morano
Language as an Ecological Phenomenon, edited by Sune Vork Steffensen, Stephen Cowley and Martin Döring
Multispecies Discourse Analysis, Gavin Lamb
Storytelling and Ecology, Anthony Nanson
TESOL and Sustainability, edited by Jason Goulah and John Katunich

Ecolinguistics, Social Justice and Sustainability

Voices from the Global South

Edited by Mohamed Mliless, Mohammed Larouz, David Stringer, Diego L. Forte, George M. Jacobs and Meng Huat Chau

BLOOMSBURY ACADEMIC
LONDON • NEW YORK • OXFORD • NEW DELHI • SYDNEY

BLOOMSBURY ACADEMIC
Bloomsbury Publishing Plc, 50 Bedford Square, London, WC1B 3DP, UK
Bloomsbury Publishing Inc, 1359 Broadway, New York, NY 10018, USA
Bloomsbury Publishing Ireland, 29 Earlsfort Terrace, Dublin 2, D02 AY28, Ireland

BLOOMSBURY, BLOOMSBURY ACADEMIC and the Diana logo are trademarks of Bloomsbury Publishing Plc

First published in Great Britain 2025

Cover design: Jade Barnett

A catalogue record for this book is available from the British Library.

A catalog record for this book is available from the Library of Congress.

ISBN: HB: 978-1-3505-2380-7
ePDF: 978-1-3505-2381-4
eBook: 978-1-3505-2382-1

Series: Bloomsbury Advances in Ecolinguistics

Typeset by Deanta Global Publishing Services, Chennai, India

BLOOMSBURY
OPEN ACCESS

An ebook edition of this book is available open access on bloomsburycolle ctions.com. Open access was funded by the Bloomsbury Open Collections Library Collective.

Bloomsbury Open Collections is a collective-action approach to funding open access books that allows select authors to publish their books open access at no cost to them. Through this model, we make open access publication available to a wider range of authors by spreading the cost across multiple organizations while providing additional benefits to participating libraries. The aim is to engage a more diverse set of authors and bring their work to a wider global audience.

More details, including how to participate and a list of contributing libraries, are available from http://www.bloomsbury.com/bloomsbury-open-collections.

Contents

Figures

Tables

Contributors

Iman Jabbar Abbas is Professor at the University of Technology and Applied Sciences, Sultanate of Oman. She holds a PhD in Teaching English to Speakers of other Languages from the University of Exeter, UK. She has an international experience in teaching English as a foreign/second language. She has several certificates in business English, teaching effectiveness, instructional design and IELTS. She has participated in numerous research projects regarding course design, instructional design and digital pedagogy. She has given several invited/plenary talks at international conferences.

Aditya Alam Soeta Bangsa gained his bachelor's degree in English Language and Education in 2021 from Universitas Lambung Mangkurat, Indonesia. He graduated from the Applied Linguistic program with a specialization in Translation Study at Universitas Negeri Yogyakarta, Indonesia. His current research interest is in tourism translation topics, including ecotourism discourse translation and cultural tourism translation.

Meng Huat Chau is Professor of Applied Linguistics and Language Education at Zhejiang International Studies University (ZISU), where he mentors early-career scholars in research and academic inquiry. He has held adjunct and visiting appointments at Jeonbuk National University, the University of Cambridge and Yogyakarta State University. Prior to joining ZISU, he dedicated over two decades to working closely with students and educators across primary, secondary and tertiary institutions. His teaching, research and mentoring focus on harnessing the study, pedagogy and practice of language and communication to foster a more inclusive and sustainable world for people, fellow animals and the environment.

Hildo Honório do Couto is Professor Emeritus at the University of Brasília, Brazil. He completed his PhD in Phonology at the University of Cologne, Germany. He began his career as phonologist and creolist. He published two books, *Creolistics: O Crioulo Português Da Guiné-Bissau* (1994) and *Introdução Ao Estudo Das Línguas Crioulas E Pidgins* (1996). As of 1998, he has been dealing only with ecolinguistics. He is the leader of a group of investigators from several Brazilian universities and some Latin American ones. He has published a number of ecolinguistics books, such as *Ecolinguística: estudo das relações entre língua e meio ambiente* (2007), the first introduction to the discipline in Portuguese. This book was followed by another one on language contact two years later, *Linguística, ecologia e ecolinguística* (2009). He co-edited the ecolinguistics anthology *O paradigma ecológico para as ciências da linguagem* (2016), which contains practically all the 'classical' essays on the subject besides some recent ones, with a total of eighteen chapters (427 pages), everything translated into Portuguese. He created an ecolinguistics website (www.ecoling.unb.br) and the journal *ECO-REBEL* (http://periodicos.unb.br/index.php/erbel/). He organizes the Brazilian Meetings on Ecolinguistics (EBE), which take place every two years. The first one was in 2012. He also has an ecolinguistics list, with 330 members. His production also includes several articles and book chapters (in Portuguese and in English) as well as the supervision of several MA and PhD theses. Finally, Hildo created the ecolinguistics theory known as ecosystemic linguistics (EL), which includes the sub-theory ecosystemic discourse analysis (EDA) based on ideas originally proposed by Peter Finke and his disciple Wilhelm Trampe, as well as on dialectical ecolinguistics formulated by the Danish ecolinguists Jørgen Bang and Jørgen Døør.

El Ayachi El Baghdady works for the Ministry of Education as secondary school teacher in Morocco. He is also a member of Applied Language and Culture Studies Lab (ALCS). He obtained his master's degree in narrative and pragmatics at the Faculty of Letters and Humanities at Chouaib Doukkali University, Morocco. He is specialized in cognitive science and currently investigating the effects of narrative on individuals' behaviour.

Diego L. Forte is Professor in Linguistics and Social Semiotics. He is currently an academic consultant for the Ministry of Agriculture of Argentina and has worked a researcher for the National Library of Argentina. He has taught Critical Discourse Analysis in the University of Buenos Aires, the National Institute of Public Administration of Argentina and postgraduate courses on multimodality, emotion and ecolinguistics in the University of La Serena, Chile.

Peter Githinji holds an MA and PhD in Linguistics from Michigan State University, USA. He taught Swahili and linguistics at Ohio University for eleven years. He teaches in the department of Kiswahili and African Languages at Kenyatta University, Kenya, where he is currently serving as the Chair. His research interests include dialectology, youth and urban languages, language contact and language and identity, and sociolinguistics in general.

Hisham M. El-Tahawy is Assistant Professor of English Literature at Arab Open University, Kingdom of Bahrain, and Head of Training, Continuous Education, and Community Service. He holds a PhD Degree in English Literature in Minia University, Egypt, the topic of campus novels and an MA in Ethnic Drama. Dr Hisham is a Fellow of the Higher Education Academy UK. He has more than twenty years of experience in academia and professional development. Currently, he is working on developing the employability skills of AOU graduates and the technological transformation during the post-pandemic period. His research interests are ethnic literature, the reflection of campus life in literary texts, African literature, ecocriticism, children's literature and educational leadership.

George M. Jacobs is an educator from Singapore. His primary research interests are student-centred instruction, cooperative learning, humane education, environmental education, teacher education and positive education. His two main roles are teaching student-centred approaches to in-service teachers in a variety of fields, and teaching in the non-formal education sector for non-profit organizations. He is also active in professional associations, such as the Extensive Reading Foundation and the International Ecolinguistics Association. George has published more than 260 articles, book chapters and books on topics related to this book. He regularly conducts workshops and courses for teachers on cooperative learning and other student-centred themes.

Gusti Agung Sri Rwa Jayantini is Professor of Linguistics at the Universitas Mahasaraswati Denpasar, Indonesia. Dr Jayantini has been awarded several research grants with national and international funding. She has published a number of textbooks as references in the field of applied linguistics, including *The Art of Translating* (2016), *Practice Makes Perfect* (2015), *Scientific Translation: Techniques and Phenomena* (2018), *Communicative English* (2021), *Drama Study* (2023) and *A Comprehensive Guide for Interpreting Practice.*

Ouarda Khouni holds a PhD in Applied Linguistics and English Language Teaching. She has been teaching English at universities and colleges for sixteen years where she has taught different subjects. She has been in charge of different duties including teaching, supervising, curriculum design, research reviewing and quality assurance. She is a member of many organizations such as IATEFL, ESOL, SRHE and ITDI, and a reviewer for a few journals. She has presented in different national and international conferences in the UK, Algeria, Oman and the UAE. Her research interests and publications include motivational strategies, anxiety, teacher quality, and web skills and tools.

Mohammed Larouz is the Dean of the School of Arts and Humanities of Meknes and Vice-President of Academic Affairs at Moulay Ismail University of Meknes, Morocco. He earned a PhD in Applied Linguistics from Fez University, Morocco, in 2004, and has worked as a Professor in the Department of English Studies since 2005. Dr Larouz also holds an MA in Applied Linguistics from Essex University, UK, and has been a Fulbright Scholar to the United States twice. He has research interests in TEFL, sociolinguistics, research methodology and communication.

Mohamed Louza is Assistant Professor of English at Cadi Ayyad University, Morocco. His research interests encompass ecocriticism, post-colonial theory and environmental justice in the context of the Global South. He has published articles on various topics, including 'The Problematic of (M)othering in Toni Morrison's Sula', 'Environmental Ethics in Jean Giono's "The Man Who Planted Trees"' and 'Narrative Scholarship in Rachel Carson's Silent Spring'.

Mouhamed Abdallah Ly is a senior researcher in Language Sciences at the Institut Fondamental d'Afrique Noire Cheikh Anta Diop (IFAN Ch. A. Diop), Senegal. His research interests include decolonial thinking and ecolinguistics. He is editor of a special issue of the journal *Les Notes Africaines*, which is devoted to ecolinguistics.

Mohamed Mliless is an independent researcher in ecolinguistics and critical environmental discourse analysis. He holds a PhD in Applied Linguistics from the University of Moulay Ismail of Meknes, Morocco. He has published books on legal and environmental discourse, and authored articles on social and environmental issues. Dr Mliless is a member of the International Ecolinguistics Association (IEA) and serves on the boards of several journals.

Hassane Razkane is Professor of Applied Linguistics and English at Chouaib Doukkali University, Morocco. He obtained his doctorate in Applied Linguistics from Chouaib Doukkali University, Morocco. He is currently exploring the cross-language transfer of (meta)linguistic skills and (meta)cognitive strategies among trilingual learners. Razkane has published a number of refereed articles on different topics related to psycholinguistics, applied linguistics and e-learning.

Mariana Roccia is a research impact consultant for the Culture, Continuity and Transformation Research Priority Area at the University of Gloucestershire, UK. She is book series co-editor of *Bloomsbury Advances in Ecolinguistics*, co-convenor of the International Ecolinguistics Association and a qualified English-Spanish translator.

Jasper Roe has nearly a decade of experience leading English and pre-university programmes in international higher education institutions in Southeast Asia, and is a current researcher in academic integrity, sociology and applied linguistics. He is the head of the Language School at James Cook University, Singapore.

Monika Christine Rohmer is Assistant of the Chair of African and Afrophone Philosophies (Prof Dr Alena Rettová) at the University of Bayreuth, Germany. She is currently pursuing a PhD on water-related discourses in Senegal in the framework of the Bayreuth International Graduate School of African Studies (BIGSAS).

Philip W. Rudd is Professor of English and Modern Languages at Pittsburgh State University, United States. As a Peace Corps volunteer in East Africa, he taught at a rural secondary school. In the 1990s, he taught on the Persian Gulf. As a doctoral student, he won a Fulbright-Hays Doctoral Dissertation Fellowship to Kenya. As a 2017–2018 Fulbright US Scholar, he researched Nairobi's urban vernacular. His research interests include ecolinguistics and language contact.

Adil Youssef Sayeh is an EFL teacher with the Moroccan Ministry of Education and a PhD candidate at the Applied Language and Culture Studies Lab (ALCS) at Chouaib Doukkali University in El Jadida, Morocco. He is interested in investigating socio-political, socio-cultural and cognitive factors related to teaching languages, and implementing multilingual approaches into education.

David Stringer is Associate Professor of Second Language Studies and Director of Undergraduate Studies, Second Language Studies, at Indiana University, United States. His main research area is the acquisition of lexical semantics and syntax, with a focus on universals of language and cognition. Additional areas of interest include language attrition, multilingualism in post-colonial societies and biocultural diversity conservation, linking indigenous language revitalization to the conservation of ecosystems. His interests extend to all formal aspects of second-language acquisition and multilingualism, and he enjoys working with students on a wide range of projects.

Sang Putu Kaler Surata is a professor, teaching Biology Education at the Faculty of Teacher Training and Education at Universitas Maha Saraswati Denpasar, Indonesia. Surata has conducted a significant amount of research on environmental issues, focusing on their integration into teaching and learning models. Having been involved in a series of internationally recognized research projects on the local values and concepts of sustainable practices, Surata has continuously worked to promote the Outstanding Universal Values (OUVs) of the *subak* system.

Ingrid Gavilán Tatin is a translator of English into Spanish, who graduated from Universidad Católica de Temuco, Chile. She holds a master's degree in Applied Linguistics with a specialization in Foreign Language Education from Universitas Negeri Yogyakarta, Indonesia. Ingrid has participated in the FEALAC Youth Declaration on Climate Actions 2023, Bandung, Indonesia. Her interests include cultural translation, ecolinguistics (especially regarding animal welfare), and the application of ecolinguistics in education.

Ronald Umbas is a lecturer, teaching English and public relations for the study program Public Administration at the Institute of Social and Political Science Wira Bhakti, Indonesia. Umbas has conducted research on translation and social issues. Having been involved in several projects for local communities, Umbas has contributed to improving the communication capabilities of small and medium business owners in their dealings with foreigners.

Mohamed Yeou is Full Professor of Linguistics and English at Chouaib Doukkali University, Morocco. He is the director of ALCS. Dr Yeou holds a doctorate in Experimental Phonetics from Sorbonne Nouvelle, France (1996). He was a Fulbright visiting professor at UCLA, United States, in 2008. Yeou has co-edited a number of books and published two books, a number of refereed articles and book chapters on different topics related to linguistics, e-learning and ELT.

Chenghao Zhu teaches at Jiangxi Science and Technology Normal University in China. He obtained his Ph.D. from Universiti Malaya and completed his master's degree in Applied Linguistics at Beijing Language and Culture University. He is a loyal fan of ecolinguistics. In recent years, the young scholar has been fortunate enough to work with outstanding ecolinguists and has joined many exciting projects, such as 'Ecolinguistics for and beyond the Sustainable Development Goals'. Chenghao's research interests include using corpus-linguistic and computational-linguistic approaches to support ecolinguistic research.

Foreword

Hildo Honório doCouto

The devastating floods that ravaged and are still ravaging the South of Brazil at the time I am writing (May 2024), among many other extreme events in various parts of the world, are convincing proof that we linguists also need to act to raise awareness, among people and governments, that life on the face of the earth is in danger. After all, we are not here and nature is there. We are part of it, not just spectators. Our relation to it is a relation of part to whole. Instead of environment, we should say something like 'co-vironment', as in the German neologism *Mitwelt* (co-world, co-vironment), sometimes used instead of the regular *Umwelt* (circum-vironment) for environment. All the chapters of this book are attempts to awaken this awareness, in one way or another. By reading them, we can glimpse the idea that nothing justifies our arrogance towards the rest of creation.

Human arrogance has probably existed since we began walking upright, leading us to consider ourselves kings of creation. For example, Protagoras of Abdera (ca. 490–420 BCE) asserted in his book *Truth* that we are the kings of creation. According to him, 'Of all things the measure is man, of the things that are, that they are, and of the things that are not, that they are not'. This has frequently been rendered as 'Man is the measure of all things'. However, as Christopher Manes appropriately said,

> If fungus, one of the 'lowliest' of forms on a humanistic scale of values, were to go extinct tomorrow, the effect on the rest of the biosphere would be catastrophic, since the health of forests depends on Mycorrhyzal fungus, and the disappearance of forests would upset the hydrology, atmosphere, and temperature of the entire globe. In contrast, if Homo sapiens disappeared, the event would go virtually unnoticed by the vast majority of Earth's life forms. (Manes 1996, p. 24)

I would like to add that if *Homo sapiens* disappears, the other species will flourish much more vigorously.

In an attempt to counter our anthropocentric view of the world, some people claim that we need to protect nature. However, nature does not need our protection. It will follow its course with us or without us. It is up to us to decide whether to go with it or perish. Thus, an apparently more reasonable position seems to be the one proposed by Arne Naess, the Norwegian philosopher who created deep ecology. In his opinion, the human being is just another living being. Human beings are not more important than any other living being, as Manes's argument also shows. According to Naess, respecting and valuing other living beings does not imply devaluing humans but inserting them into a larger context, that of life on the face of the earth. We are important and have the right to live, but members of other species are also important and have the right to live. It is true that in nature there are food chains. Very often one being eats another or kills it to defend its own integrity and/or territory. Even we humans defend ourselves against some beings that can harm our existence. All of this exists to maintain balance in living nature (Naess 1989).

This book is an important step in the direction of showing that if we want to remain alive, we must live harmoniously with other living beings as well as with their (and our) physical basis such as soil, water, air and so on. The essays contained in this book generally use as their foundation the branch of linguistics whose name begins with *eco-*, namely ecolinguistics, from *oikós* in Greek, where it means house. The book's chapters invite and encourage us to have a non-predatory relationship with the world outside our bodies. They analyse the way we relate to the environment by observing the language used in this relationship. The texts show that we must take care of our house-environment because if it collapses, we collapse with it.

I will not present all ten chapters of the book because this is already well done in the Introduction and in Chapter 1 by Mariana Roccia. It is more important to point out that the diversity of provenience of the authors and of the subjects they deal with is in tune with one of the central concepts of ecology and diversity. The more diversity of life there is in an ecosystem, the richer and more vibrant it will be. Ecology is part of biology, the science of life. Therefore, it is no coincidence that practically every chapter talks about life, directly or indirectly.

The countries in which the authors work are not necessarily their countries of origin, which increases the diversity of perspectives. Here is the list of countries in which they are working, in alphabetical order:

- Argentina: Diego L. Forte
- Germany: Monika Christine Rohmer

- Indonesia: Gusti Agung Sri Rwa Jayantini, Sang Putu Kaler Surata, Ronald Umbas, Ingrid Gavilán Tatin, Aditya Alam Soeta Bangsa
- Kenya: Peter Githinji
- Kingdom of Bahrain: Hisham M. El-Tahawy
- Malaysia: Chenghao Zhu, Meng Huat Chau
- Morocco: Hassane Razkane, El Ayachi El Baghdady, Adil Youssef Sayeh, Mohamed Yeou, Mohamed Mliless, Mohammed Larouz, Mohamed Louza
- Senegal: Mouhamed Abdallah Ly
- Singapore: George M. Jacobs, Jasper Roe
- Sultanate of Oman: Iman Jabbar Abbas, Ouarda Khouni
- United Kingdom: Mariana Roccia
- United States: Philip W. Rudd, David Stringer

The first chapter, by Mariana Roccia, is an excellent overview of ecolinguistics, including its manifestations in Latin America. The other nine chapters investigate: (Chapter 2) a study carried out with school children in Morocco by Hassane Razkane, El Ayachi El Baghdady, Adil Youssef Sayeh and Mohamed Yeou; (Chapter 3) the integration of ecolinguistics into the EFL curriculum in Oman; (Chapter 4) how children's literature can foster critical thinking about environmental issues; (Chapter 5) the Outstanding Universal Values of *subak*, emphasizing 'the importance of preserving cultural landscapes and the philosophy of *Tri Hita Karana*' in Bali; (Chapter 6) the promotion of ecotourism in Morocco through YouTube videos; (Chapter 7) the discourses of Argentina's food production control agency, SENASA, and their ties to colonial, eco-industrial and speciesist ideologies; (Chapter 8) ecological identity in Kenya through Swahili lexicogrammar, underscoring 'the need for identities that benefit ecosystems amidst increasing climate change' through 'Swahili newspaper articles on climate change'; (Chapter 9) decolonial ecolinguistics through marginal biographies, focusing on Ibrahima Diop, a Senegalese man living in Dakar's peri-urban forests; and (Chapter 10) the impact of civet coffee production on civets in Indonesia.

In her overview of ecolinguistics in Latin America, Roccia mentions its Brazilian branch, ecosystemic linguistics (EL), as well as its sub-branch ecosystemic discourse analysis (EDA). Since these theories are little known but their principles are entirely compatible with the content of all chapters, I would like to briefly present them. They are not like recipes for making cakes nor like generative grammar's model for the study of syntax but a unified point of view from which you can study any and all aspects of language (holism) using the ecomethodology known as the 'focussing method', proposed by Garner (2004). Thus, if you do not have the necessary knowledge to investigate a certain subject, such as cleft sentences in syntax, the status of nasal vowels in phonology or the hidden meanings of a discourse, you can ask an expert to carry out the analysis, whose results you can evaluate from the holistic perspective of EL and EDA. This perspective is the ecological view of the world, in contrast to the Occidental view of the world. What matters most is evaluation, not the analysis per se.

Ecosystemic linguistics is so called because it departs from the biological ecosystem to build its theoretical foundations. The ecosystem is the setting in which the drama of life unfolds. EL is radically ecological because it is part of biological ecology, not by using ecological concepts metaphorically like most ecolinguistics theories. It is linguistic ecology, not ecological linguistics. Thus, whereas the biological ecosystem consists of the interactions (I) between members of a population (P) of organisms as well as between them and their habitat or territory (T), the linguistic ecosystem is formed by a people (P) in its territory (T) and the word-centred interactions that are obtained therein, namely, the language (L). To put it differently, language is not simply an instrument of communication. It is communication, or communicative interaction. This can be seen in the fact that the L of the linguistic ecosystem is equal to the I of the biological ecosystem. In terms of application, the initial triad LPT is a strong argument in favour of minority groups' integrity and identity. The loss of L is a first step towards the annihilation of the group. The loss of T is a definitive step towards its disappearance.

LPT is the integral ecosystem of language. It contains three sub-ecosystems in its interior, namely, the natural, the mental and the social ecosystem of language. Phenomena of the natural ecosystem of language are studied by phonetics, paralinguistics, proxemics and kinesics, among others. Phenomena of the social ecosystem of language are studied by sociolinguistics, ecosystemic discourse analysis and so on. The mental ecosystem of language is the mediator between the two. In it, everything linguistic is processed. Generative grammar sees language exclusively from this perspective.

The sub-branch of EL known as EDA (ecosystemic discourse analysis) does not emphasize ideologies – frequently Marxist ideology – and power relations like traditional discourse analysis. Its two fundamental principles are: (1) unconditional defence of life; and (2) the fight against avoidable suffering. Inevitable suffering/pain also exists but in order to lead organisms to avoid the mutilation of their bodies. Since EDA looks at discourses holistically, political and other ideologies are not ignored, but they are subordinated to the ideology of life, ecoideology/bioideology.

When you cross the threshold of the integral ecosystem of language, you enter a world in which everything will be seen in a way that is radically different from the Western view of the world. Here, we could remember Dante Alighieri's famous verse from the first part of his *Divina Commedia*: 'Abandon all hope you who enter'. Ecosystemic-linguistically this is a warning not to see linguistic phenomena from the perspective of the Western worldview but rather to see them from the ecological worldview. This is not easy because we do not like leaving our comfort zone.

Ecosystemic linguistics and ecosystemic discourse analysis are extensively and intensively discussed in the journal *ECO-REBEL*, available online, as well as in other publications like those mentioned by Mariana Roccia in this book. Unfortunately, the two theories are discussed mainly in Portuguese, but there are some publications in English as well, as is the case with EL in Couto (2018) and with EDA in Couto.

I consider this brief outline of EL and EDA pertinent in this book because its chapters are geared towards problems of the Global South, where EL and EDA emerged, and because the subjects they deal with have the same ultimate aim as these theories. It is possible to see this even in the outlines of the first two chapters. Chapters 3 through 10 show this even more convincingly. As already pointed out above, most of them deal with issues that have to do with the threats to life on the face of the earth. As a matter of fact, in one way or another, most ecolinguistic investigations around the world aim at defending life on earth, explicitly or implicitly. In the paragraph following the one that lists the authors and the countries in which they work, I have presented the list of these subjects.

In regard to the idea of the Global South, I would like to say that the Brazilian group that practices EL and EDA defends the right and duty to produce science also in Latin America's local languages, Portuguese and Spanish, without ignoring the international multilingual panorama, which is generally expressed in English. It is not by chance that perhaps over 80 per cent of the publications in EL and EDA are in Portuguese. However, there are some offshoots of these theories in Latin American countries like Argentina, Bolivia and Chile. Two of the most relevant publications in Portuguese are the 2016 anthology mentioned in Albuquerque's chapter and the journal *ECO-REBEL*, the organ of the Brazilian ecolinguistic movement, although it has published many articles in Spanish and English as well. As far as I know, there is nothing similar in Spanish, French or Italian. In summary, Latin America and the Global South at large also produce science in non-English languages. It would certainly be very enriching if a dialogue emerged between these South American movements and the authors of the different chapters of the book, not to mention its potential readers.

Last but not least, I recommend the book *Ecolinguistics, Social Justice and Sustainability: Voices from the Global South* for at least three main reasons. First, its essays suggest that we interact with the world external to our bodies in a sustainable and harmonious way. Second, they also defend harmonious interaction at the social level, directly or indirectly. Third, they show the importance of ecolinguistics in all these processes, despite the fact that it is seen as peripheral and marginal by practitioners of mainstream linguistics in academia. However, we ecolinguists don't care what 'status quo' linguistics thinks about us. We are convinced that we are on the right path in the face of the dangers of life disappearing from the face of the earth. After all, language is intimately linked to life.

References

Couto, H. H. (2018). Ecosystemic linguistics. In A. Fill & H. Penz (Eds.), *The Routledge Handbook of Ecolinguistics* (pp. 149-161). London: Routledge.

Couto, H. H., Couto, E. N. N., Silva, A. N. (2021) Ecosystemic Discourse Analysis (EDA). *ECO-REBEL* v. 7, n. 1, pp. 05-17, available at https://periodicos.unb.br/index.php/erbel/article/view/36017/29022

Manes, C. (1996). Nature and silence. In: Glotfelty, Ch., Fromm, H. (orgs.). *Ecocriticism reader: Landmarks in Literature and Ecology*. Athens: The University of Georgia Press, pp. 15-29.

Naess, A. (1989). *Ecology, community and lifestyle*. Cambridge: Cambridge University Press.

Acknowledgement

This publication is the fruit of a collaborative endeavour, brought to fruition by our esteemed editors Dr Mohamed Mliless, Dr Mohammed Larouz, Dr David Stringer, Dr Diego Luis Forte, Dr George M. Jacobs, and Dr Meng Huat Chau. Our heartfelt gratitude extends to all contributors whose unwavering diligence has been instrumental in the timely completion of this work.

We would also like to express our profound gratitude to Dr Hildo Honório do Couto, who generously contributed his time and expertise to write a preface for our book. His thoughtful insights and perspectives have added significant depth to our work, and his support has been instrumental in bringing this project to fruition. We are deeply appreciative of his contribution and the impact it has made on our collective endeavour. Thank you, Dr Honório, for your invaluable assistance and for being a part of this journey with us.

We extend our special appreciation to the eminent professors who graciously provided forewords for this book. Their invaluable insights have enriched the content and underscored the significance of numerous chapters. We are deeply thankful for their time and effort, which have significantly amplified the overarching narrative of the book.

Every effort has been made to trace copyright holders and to obtain their permission for the use of copyright material. However, if any have been inadvertently overlooked, the publishers will be pleased, if notified of any omissions, to make the necessary arrangements at the first opportunity.

Abbreviations

ACS	AfricaClimate Summit
ANIS	Articulation of narratives and images on screen
AOF	Afrique occidentale française
CDA	Critical discourse analysis
COP	UN Conference of the Parties
CTL	Contextual teaching and learning
EAP	English for academic purposes
EDA	Ecological discourse analysis
EE	Environmental education
EFL	English as a foreign language
ELT	English-language teaching
ES	Education for sustainability
ESL	English as a second language
GACC	Administración General de Aduanas de China
GACC	Nation and the General Administration of Customs of China
GCE	Global citizenship education
GDP	Gross domestic product
CGEM	General Confederation of Moroccan Enterprises
HAD	Harmonious discourse analysis
IASCAV	Argentine Institute for Vegetal Health and Quality
IDR	Indonesian rupiah

JP	*Jakarta Post*
KP	*Kompas*
MAD	Moroccan dirham
MMC	Mental model category
NEEF	National Environmental Education Foundation
NiE	Newspapers in Education Programme
NMG	Nation Media Group
OUVs	Outstanding Universal Values
PES	Plan for an Emerging Senegal
PETA	People for the Ethical Treatment of Animals
SARS	Severe acute respiratory syndrome
SDGs	Sustainable Development Goals
SENASA	National Agri-Food Health and Quality Service
SEST	Socio-ecological systems theory
SFL	Systemic-functional linguistics
SNS	Social network sites
UAE	United Arab Emirates
UNEP	United Nations Environment Programme
UNESCO	United Nations Educational, Scientific and Cultural Organization
UNWTO	United Nations World Tourism Organization
UTAS	University of Technology and Applied Science
VYI	Foreign YouTube influencers

Introduction

Mohamed Mliless

Ecolinguistics, Social Justice and Sustainability: Voices from the Global South is a collection of chapters that delves into various interests and fields of research associated with ecolinguistics. On the website of the International Ecolinguistics Association, ecolinguistics is defined as having two aims in exploring,

> the role of language in the life-sustaining interactions of humans, other species and the physical environment. The first aim is to develop linguistic theories which see humans not only as part of society, but also as part of the larger ecosystems that life depends on. The second aim is to show how linguistics can be used to address key ecological issues, from climate change and biodiversity loss to environmental justice.

For an understanding of the roots of ecolinguistics, please see the comprehensive overview provided by Fill and Penz in their 2018 handbook on ecolinguistics. Additionally, Steffensen and Fill (2014) and Stibbe (2021) are leading authors who address the future of the field. Many scholars trace the origins of the work found in this book and in much other work in ecolinguistics to M.A.K. Halliday's talk at the 1990 AILA conference in Thessaloniki, Greece, titled 'New Ways of Meaning: The Challenge to Applied Linguistics', republished in 2001. In his talk, Halliday emphasized how language shapes people's thinking and actions, giving specific examples of what he labelled as growthism, the view that more and more products need to be produced and sold, thereby more demand must be created. Of course, all these products require resources which must be ripped out of the Earth.

This book provides examples of how language and ecology meet in the Global South. This entails investigating a wide range of themes and interactions in the setting of the Global South. Language, as Halliday explained, has a significant impact on our perceptions and knowledge of the world around us. *Ecolinguistics, Social Justice and Sustainability: Voices from the Global South* presents both theory and empirical research findings in the field of ecolinguistics.

The book is intended for students and professionals who wish to gain a broader knowledge of how language plays a strategic role in tackling ecological concerns at various societal levels. The chapters of the book relate to the overall theme by exploring various aspects of the intersection between language and ecology in the context of the Global South. Each chapter focuses on a specific topic within this broad theme, such as the role of language in shaping our understanding of the natural world, the impact of colonialism and globalization on language and ecology, and the practical applications of ecolinguistic research. By presenting a diverse range of perspectives and approaches, the chapters collectively provide insights into the field of ecolinguistics and its relevance to the Global South.

The book's preface

This book's preface by Hildo Honório do Couto reminds us that as humans, 'we are not here and nature is there. We are part of it, not just spectators. Our relation to it is a relation of part to whole'. Humans, he tells us, must abandon the arrogance with which for millennia we have looked down on the rest of our planet's flora and fauna. We should be reminded that, for example, life as we know it would cease to exist without the lowly fungi, and without humans, much of Earth's other inhabitants would thrive. Honório do Couto also introduces two lesser-known theories in ecolinguistics: ecosystemic linguistics and its sub-branch ecosystemic discourse analysis. The online journal *ECO-REBEL* is one place to learn more about these theories. It is also a place where another concern of Honório do Couto's is addressed, that too much of ecolinguistics goes on in English. The preface concludes with a three-fold hope for what the book's chapters could inspire. First, the chapters call for harmony in human interactions with the non-human world. Second, the chapters also support harmonious interaction among humans. Third, the chapters demonstrate the need to use ecolinguistics as a tool in order to achieve this harmony, 'despite the fact that it is seen as peripheral and marginal by practitioners of mainstream linguistics in academia. However, we ecolinguists don't care what "status quo" linguistics thinks about us'.

Themes of the book

Promotion of ecolinguistics in education

Given the significant role of language in shaping human perceptions and behaviours, it is crucial to integrate ecolinguistics into language teaching curricula at all levels of education. This can be achieved through the use of ecologically conscious texts and activities that foster ecoliteracy. By doing so, we can help students develop a more nuanced and environmentally aware understanding of language and its relationship to the natural world. In the Latin American context, Chapter 1 by Mariana Roccia reviews the progress, challenges and opportunities for ecolinguistics in Latin America. She posits that, despite global advancements, ecolinguistic research in Latin America needs further development. The chapter aims to stimulate scholarly conversation and research for local and practical responses relevant to the region. Latin America, home to many indigenous communities and coveted natural resources, has a long history of environmental, economic and social problems often interlinked with its colonial past and with ongoing exploitation by forces from the Global North. This raises the question of why there has been little interest from scholars in Latin America in exploring the field of ecolinguistics more actively. Can some of the challenges stem from the predominance of the English language in the production and distribution of the literature? Is it possible that research that fits the 'ecolinguistic' criteria has not been labelled accordingly? To address these and other concerns, this chapter first offers a review of the advances in Latin America while reflecting on current challenges that ecolinguists face. The analysis enables a more nuanced understanding of the nature of the ecolinguistics research that can be applied in the Latin American context. Finally, and more ambitiously, the chapter aims to spark scholars' interest in engaging in conversations and producing research that builds towards the necessary local and practical responses relevant to this region and its communities.

Holistic environmental education

The observed gap between environmental education and the environmental practices of students, teachers and society suggests an urgent need for more holistic approaches to environmental education. These approaches should focus on fostering empathy, responsibility and agency among learners. By moving beyond traditional educational models, we can create more meaningful and transformative learning experiences that inspire real-world environmental action. In this dimension, the book continues with Chapter 2 by Hassane Razkane, El Ayachi El Baghdady, Adil Youssef Sayeh and Mohamed Yeou, investigating the gap between environmental education and students' environmental practices. While environmental education is widely acknowledged as a crucial means to promote sustainable behaviours, a noticeable gap often exists between the environmental knowledge students receive at school and their practices in real-life environmental settings. This chapter examines students' undesirable environmental behaviours despite receiving education on environmental protection. Drawing on socio-ecological systems theory (Berkes & Folke, 1998; Ostrom, 2009), which examines how social influences, psychological motives and institutional dynamics collectively shape individuals' environmental behaviours, this study explores the underlying factors behind this discrepancy between the curriculum and students' actions towards the environment. To investigate middle school students' perceptions and motivations towards environmental practices, data were collected via a questionnaire and semi-structured interviews. Descriptive statistics were used to analyse the questionnaire data, and thematic analysis was employed to explore the interview data. The study's findings suggest a complex interaction of factors that affect students' environmental habits. These include psychological barriers, such as students' internal contradictions between their awareness of environmental issues and their everyday practices towards the environment, and social influences, including peer pressure and conflicting cultural norms. The implications of the study point towards the necessity of holistic approaches to environmental education. The authors call for pedagogical strategies that not only deliver knowledge but also actively cultivate a sense of responsibility, agency and empathy towards the environment and promote generations of environmentally conscious citizens.

Integrating ecolinguistics into the EFL curriculum

There is a growing interest in incorporating ecolinguistics into different educational environments as people become more conscious of the linkages between language, culture and the environment. With a focus on how linguistic practices might support ecological sustainability, ecolinguistics studies the interaction between language and the natural world. This multidisciplinary topic provides insightful information about how language instruction might encourage sustainable habits and environmental awareness. To explain this nexus, Chapter 3 by Iman Jabbar Abbas and Ouarda Khouni discusses the integration of ecolinguistics into the EFL curriculum in Oman. Their study explored the impact of ecolinguistics integration on the motivation from the perspectives of foundation-level students at the University of Technology and Applied Sciences in Oman. The study also investigated the strategies and methods of integrating ecolinguistics into the EFL curriculum in the higher education context in Oman. The study employed a questionnaire survey as a method of collecting the data quantitatively. The findings of the study revealed that the respondents were in favour of integrating ecolinguistics into the existing curriculum, and the process of integration could be done via an effective selection of materials and activities. The study contributes to knowledge by providing a synthesis of theoretical and practical views regarding the integration of ecolinguistics in the EFL curriculum.

Critical thinking through children's literature

The potential of children's literature in sparking critical thinking about environmental issues cannot be overstated. We recommend incorporating more environmentally themed literature into children's education as well as broader societal contexts to effectively foster critical thinking about ecological conservation. Of importance to this theme, Chapter 4, authored by Chenghao Zhu, Meng Huat Chau, Hisham M. El-Tahawy, Jasper Roe, and George M. Jacobs, explores how children's literature can foster critical thinking about environmental issues. Children's books provide children and others, including learners of additional languages, with invaluable opportunities for both enjoyment and education. The chapter offers ideas on how children's literature can also be a vehicle for promoting critical thinking, in particular about the current environmental crisis that humans are coping with. To promote such critical thinking, concepts from ecolinguistics are described. These ecolinguistics concepts include the nine types of 'stories we live by', that is, ways of viewing the world and deciding what behaviours to enact in the world (Stibbe, 2021). A children's book, *Thirst*, by Varsha Bajaj, is summarized and used for examples of some concepts from ecolinguistics as well as goals from the United Nations' Sustainable Development Goals (SDGs). The SDGs are suggested as ideas that show the interconnectedness of struggles to make the world a better place for all people, regardless of background. The authors hope to encourage greater appreciation of children's literature and greater use of this literature to spark critical thinking about, concern for and action on behalf of our planet's ecosystems, and the billions of humans and other animals suffering unjustly from environmental damage.

Fostering ecocultural awareness

The integration of the Outstanding Universal Values of *subak* into English learning demonstrates the profound importance of preserving cultural landscapes and philosophies. We recommend incorporating local cultural values and practices into language learning to foster ecocultural awareness and practices. This approach not only enriches language education but also helps preserve and celebrate diverse cultural approaches to environmental stewardship. In terms of ecocultural awareness, Chapter 5, by Gusti Agung Sri Rwa Jayantini, Sang Putu Kaler Surata and Ronald Umbas, discusses the importance of preserving cultural landscapes. In 2012, the cultural landscape of Indonesia's Bali province was inscribed by UNESCO as a World Heritage Site – a place of 'outstanding universal value', to be protected and preserved for all humankind. UNESCO highlighted Bali's *subaks*, which date back at least 1,000 years. *Subaks* are organizations of farmers who work together to manage irrigation systems for rice terraces and water temples. Farmers still use the *subak* system. The system is based on three causes of goodness: harmony in the spiritual realm, harmony between humans and nature, and harmony among humans. The chapter explores integrating the three causes of goodness into the activities of English learning, specifically at the secondary education level. This level is chosen in consideration of the crucial role of Bali's young generation to protect their cultural landscape, which is not only the land as physical objects, such as rice terraces that people may enjoy, but also the Balinese Hindus' philosophy upon which the *subak* system is based. The young generation must be reminded about their obligation to preserve the ancestors' legacy through the harmony with God, harmony with nature, and harmony with other human beings, which must all be done for sustainable living. In this chapter, the authors utilize a mental model analysis through four main activities: (1) students drawing pictures to represent their natural surroundings, including the rice fields, (2) listing the English vocabulary related to their writing, (3) interviewing the students to dive into their opinions about the importance of *subak* and protecting the green atmosphere in English learning, and (4) analysing all the gathered data from the drawings, vocabulary and interviews to map the mental model and ecocultural-based English teaching materials. Based on these data, the authors offer suggestions on how to devise English learning activities that integrate ecolinguistic perspectives.

Ecolinguistics and ecotourism promotion

The promotion of ecotourism through digital platforms, particularly YouTube, represents a significant intersection of language, culture and environmental sustainability. As influencers and content creators share their travel experiences, they wield considerable power in shaping public perceptions of ecotourism and its values. The language used in these videos not only conveys information but also constructs narratives that can either reinforce or challenge existing environmental practices and beliefs. By examining the linguistic choices made by influencers, we can gain insights into how these narratives contribute to broader discussions about sustainable tourism and its impact on local communities and ecosystems. To explain this intersection, Mohamed Mliless, Mohammed Larouz and Mohamed Louza, authors of Chapter 6, analyse the promotion of ecotourism through YouTube videos. Using ecolinguistics, Chapter 6 discusses the language used in the videos to account for the concept of sustainable tourism, which aims to minimize negative impacts on local communities and the environment. From an ecolinguistic perspective, the chapter attempts to argue that ecotourism is a form of sustainable development that focuses on preserving natural areas and wildlife while providing educational experiences for travellers. Importantly, the chapter, through the lens of ecolinguistics, argues that videos posted on YouTube by influencers about places they visited in Morocco play a key role in promoting ecotourism by communicating its principles and benefits, showcasing natural landscapes and sharing practical information on remote places in Morocco. The authors of this chapter found significant discrepancies between the content of the videos posted by foreign influencers and those by the Moroccan National Tourism Office in terms of sustainability, resilience of natural elements and the presence of human beings. The chapter suggests that the language used by influencers through videos they post on YouTube is more effective in engaging audiences and promoting ecotourism than traditional media strategies.

Ecolinguistics: Unlocking ethical insights in food production

There is a critical need for consumers, producers and policymakers to work together to ensure the welfare of all living beings and sustainability in our practices. By prioritizing ethical considerations, we can create more just and environmentally sound systems. In this direction, Chapter 7 by Diego L. Forte examines the discourses of Argentina's food production control agency, the

National Agri-Food Health and Quality Service, known as SENASA for its initials in Spanish (Servicio Nacional de Sanidad y Calidad Agroalimentaria). The discourses' ties to colonial, eco-industrial and speciesist ideologies are explored. The agency describes its main duties as to 'plan, organize and execute specific programs and plans that regulate production, directing it towards obtaining safe food for human and animal consumption'. Nevertheless, the discourses it produces for the public introduce elements that conflict with that affirmation. In earlier studies, the author had pointed out that coloniality is at the root of Argentina's national identity, and it is materialized through different kinds of texts (Forte, 2019). In this chapter, he evidences that SENASA's discourses are based on a neoclassical economics narrative that articulates colonial, eco-industrial and speciesist meanings (Hribal, 2007). Following this affirmation, the chapter analyses the construction of this narrative in two audio-visuals issued by SENASA for the general public. The analysis adopts ecolinguistics as the main theoretical framework for eco-narratives (Stibbe, 2021) and Lim and Tan-Chia (2023) for multimodal transposition. Analysis was developed through the analytical tools presented by Kress and van Leeuwen (2005) for image analysis, Ekman and Friesen (1975) for facial expression, and Forte (2023) for music analysis.

Eco-Friendly identities in Global South communities amidst climate change

Climate change is intensifying globally, highlighting the need for eco-friendly identities in communities, especially in the Global South. These communities rely on natural ecosystems for livelihoods, and climate change threatens their sustainability. Cultivating eco-friendly identities helps adapt to changing conditions, enhance resilience and contribute to global efforts to mitigate climate change. This includes integrating traditional knowledge with modern conservation practices. In Chapter 8, Philip W. Rudd and Peter Githinji delve into ecological identity in Kenya through Swahili lexicogrammar. As climate change worsens, the urge to foster identities beneficial to ecosystems has become widespread globally. A beneficial ecological identity introduces a sense of interrelatedness with other species as part of life's wider community. News media frequently grow readership and sales through marketing to create interest and model good citizenship. In fact, *Taifa Leo*, Kenya's oldest Swahili periodical, participates in the Newspapers in Education Programme (NiE), an effort to promote a culture of reading and an awareness of environment in Kenya. By analysing ecological identity, this chapter combines methods from systemic-functional linguistics (SFL), ecolinguistics and critical discourse analysis. The data set consists of ten purposefully sampled Swahili newspaper articles, containing the phrase *mabadiliko ya tabianchi* 'climate change', occurring between the UN Conferences of the Parties COP25 in 2019 and COP28 in 2023. Anthropocentrism, an ecological stance dangerous to nature, was found in the articles. The authors discuss how articles frame climate change to stimulate interest and marginalize nature as merely existing to serve human needs and desires, such as entertainment. They also supplement their quantitative analysis with a framing of cultural codes to chronicle how the issues of decoloniality, development and tradition model a more ecologically responsible identity.

Inclusion of marginalized voices

The exploration of decolonial ecolinguistics through marginal biographies underscores the paramount importance of including marginalized voices in academic discourse and policy-making. Concerted efforts must be made to ensure that diverse perspectives – particularly those from the Global South – are represented and valued in ecolinguistic research. Marginalized voices are being heard in Chapter 9 by Monika Christine Rohmer and Mouhamed Abdallah Ly, who explore decolonial ecolinguistics through marginal biographies, focusing on Ibrahima Diop, a Senegalese man living in the peri-urban forests of Dakar (Senegal's capital). Diop, for a long time, had the reputation of being a criminal or 'bandit'. This reputation was bestowed on him by those seeing him for many years in the company of a pack of dogs, heading through the peri-urban forests. These forests and non-urbanized lands within the city constitute a refuge for many marginalized people, including delinquents. It has only been recently, thanks to social media, that the general public has discovered a relatively unknown facet of Diop: he grows vegetables for his own needs, tames anacondas and experiments with plant-based healing methods using only indigenous ecological knowledge. The authors propose Diop's way of life as a form of decolonial resistance by applying the theoretical reflections of Arturo Escobar (2016, 2018) and Malcom Ferdinand (2019, 2021). They suggest that the unique relationship of Diop with his environment invites ecological reflection and rethinking. His way of life and self-narration counter current trends in Senegal and other countries in the Global South that follow a developmentalist agenda and ideology at the expense of indigenous knowledges. Thus, this chapter is in line with recent academic endeavours to learn from biographical accounts of individuals, especially those situated on the margins of society (Daha, 2002; Faye & Thioub, 2003). The chapter's authors give voice to people often disregarded in traditional academic approaches or in surveys designed for public policy purposes. They advocate for the inclusion of marginal biographies and voices in ecolinguistics. They show how life stories on the margins can contribute to decolonizing environmental knowledge. At the same time, they shed light on the entanglements of marginal lives in contemporary modernity. The authors also report limitations to interpreting marginal African biographies from a decolonial perspective. Their chapter is part of a larger effort that questions globalized discourses through the analysis of texts in African languages (Ly & Rohmer, 2024). It is a contribution at the edge of linguistics and language science. The authors are looking at the 'marginalized' biography of Ibrahima Diop, narrated in a 'marginalized' language, namely Wolof, and reflecting on the theoretical implications of this narration for decolonial ecolinguistics.

Language and environmental justice

The case of civet coffee production serves as a stark reminder of the need for responsible and ethical practices in all aspects of our lives, including food production. In this regard and in relation to the intersection of environmental justice and language, the book concludes with Chapter 10 by Ingrid Gavilán Tatin, George M. Jacobs, Meng Huat Chau, Chenghao Zhu and Aditya Alam Soeta Bangsa. This chapter highlights the importance of responsible and ethical practices in the relationship between humans and other animals, using civet coffee production as a case study. Civets, small mammals native to tropical regions of Africa and Asia, play a central role in this process. Civet coffee, known as *kopi luwak* in Indonesia, is a premium, high-priced gourmet coffee produced in Indonesia and other countries. It is made from coffee beans, also called coffee cherries or berries, that have been eaten and digested by civets before being collected and processed. Unfortunately, the attention given to civet coffee and the coffee's profit-making potential have resulted in harm to the civets. In order to increase production of civet coffee, civets are held captive and may be fed only coffee beans even though coffee is not a significant part of their normal diets. Thus, a factory farming model, long used with chickens and other farmed animals, is being employed to expand the production of civet coffee. The chapter looks at articles from two leading newspapers in Indonesia. The articles were in Bahasa Indonesia (Indonesia's national language) and English. These texts were examined using tools from the ecolinguistics literature, including Stibbe's (2021) nine stories.

This chapter fills a gap in the ecolinguistic literature by providing another example of a farmed animal and how these other-than-human beings are represented. Findings reveal that the language and practices surrounding civet coffee have both similarities to and differences from the situations of other non-human animals who are exploited to produce food and beverages for human animals.

Conclusion

To conclude this Introduction, let us return to the definition of ecolinguistics offered by the International Ecolinguistics Association, in particular to the definition's final two words, 'environmental justice'. This concept demands that attention be given to the other-than-human beings whose habitats are being destroyed and who have been kidnapped from birth and denied any semblance of a normal life before being slaughtered while still young. The call for environmental justice must also be heard in terms of humans. The irony is that those humans who contribute least to the environmental harms of the Anthropocene are the ones who suffer most, as they often have the least protection from heat, droughts, floods, storms, pollution and so on, critical issues sown mostly in the Global North. Bringing the issues of the Global South to the fore is why this book, *Ecolinguistics, Social Justice and Sustainability: Voices from the Global South*, needs to be read, to be learned from, and to be acted upon.

References

Berkes, F., & Folke, C. (Eds.). (1998). *Linking social and ecological systems: Management practices and social mechanisms for building resilience*. Cambridge University Press.

Daha, Ch. (2002). *Marginalité et exclusion au Sénégal: les comportements delictuels et criminels dans la vallée du fleuve Sénégal (1810–1970)*. [Doctoral dissertation, Université Cheikh Anta Diop de Dakar].

Ekman, P., & Friesen, W. V. (1975). *Unmasking the face: A guide to recognizing emotions from facial clues*. Prentice-Hall.

Faye, O., & Thioub, I. (2003). Les marginaux et l'État à Dakar. *Le Mouvement Social, 3*(204), 93–108.

Ferdinand, M. (2019). *Une écologie décoloniale: Penser l'écologie depuis le monde caribéen*. Seuil.

Ferdinand, M. (2021). *Decolonial ecology: Thinking from the Caribbean world*. John Wiley & Sons.

Fill, A., & Penz, H. (Eds.). (2018). *The handbook of ecolinguistics*. Routledge. https://doi.org/10.4324/9781315625296

Forte, D. L. (2019). *La construcción de discurso identitario nacional argentino: La historia de la carne*. SEDICI, Universidad Nacional de La Plata. http://sedici.unlp.edu.ar/handle/10915/94339

Forte, D. (2023). Music and discourse: A systemic-functional approach for music analysis in multimodal contexts. *Multimodality & Society, 3*(1), 69–81. https://doi.org/10.1177/26349795231153963

Halliday, M. A. K. (2001). New ways of meaning: The challenge to applied linguistics. In J. Webster (Ed.), *On language and linguistics* (pp. 139–174). Continuum. (Original work published 1990).

Hribal, J. (2007). Animals, agency, and class: Writing the history of animals from below. *Human Ecology Review, 14*(1), 101–112.

Lim, F. V., & Tan-Chia, L. (2023). *Designing learning for multimodal literacy*. Routledge.

Ly, M. A., & Rohmer, M. C. (2024). *Présentation du Numéro des Notes Africainessur : Regards interdisciplinaires sur la recherche écolinguistique au Sénégal*. Notes Africaines.

Ostrom, E. (2009). *Au-delà des marchés et des États : la gouvernance polycentrique des systèmes économiques complexes* [Beyond markets and states: The polycentric governance of complex economic systems]. Retrieved from: https://journals.openedition.org/lectures/41778

Steffensen, S. V., & Fill, A. (2014). Ecolinguistics: The state of the art and future horizons. *Language Sciences, 41*, 6–25. https://doi.org/10.1016/j.langsci.2013.08.003

Stibbe, A. (2021). *Ecolinguistics: Language, ecology, and the stories we live by* (2nd ed.). Routledge. https://doi.org/10.4324/9781000293654

1

Advances, challenges and opportunities for ecolinguistics in Latin America

Mariana Roccia

Introduction

Why Latin America?

Latin America's historical reliance on its 'natural resources' has often brought political, social and economic turmoil (Coletta & Raftopoulos, 2016) – a curse stemming from colonial years to the present (Haarstad, 2012). While the trend over the years has moved from progressive to more neoliberal governments, wealth distribution remains problematic. Models focusing obsessively on growth and development increase inequality and uncertainty among the population. For some, development is a zombie category (Gudynas, 2011), a project that is still heavily critiqued but with failed attempts at producing alternatives. Counteracting these discourses can be challenging; however, more radical social movements and indigenous communities have been given a more public voice to share their knowledge and offer new ways of revisiting concepts such as growth and development, mostly thanks to the emergence of the new left or progressive political movements. But the critique of the Eurocentric paradigm 'does not refer to displacing European thought, but rather to denying the Eurocentric vision of the world and the way of telling stories. This criticism requires a narrative from our own plural visions and versions of the world' (Corbetta, 2021, p. 111). Therefore, the search for new stories in Latin America will undoubtedly have a different aim than in the Global North. While there could be some similarities in core values, the lived experience and the historic meanings associated with the natural environment, alongside socio-economic struggles, will put the focus in a different direction.

One well-documented concept is *Buen Vivir*, an umbrella term that encompasses 'the broad notion of well-being and cohabitation with others and Nature' (Gudynas, 2011, p. 441), a concept that detaches itself from dualistic notions of linear expansion and is context-bound to different indigenous communities. Thus, it challenges Euro-modernity and its conceptualization of nature as a resource, as in the case of Latin America taking a more post-structuralist and post-socialist stance. Most importantly, Ecuador and Bolivia have included the concept of Buen Vivir in their constitutions, which represents a significant shift in perceptions of the relationship with the more-than-human and, crucially, reinstates the critical contribution of indigenous knowledge in the political domain. In practice, however, the impossibility of translating specific terms remains: 'in the cases of both Ecuador and Bolivia, the contradictions inherent in the attempts to turn constitutional principles into policy are apparent first and foremost in the abundance of modernist linguistic concepts such as "growth", "productivity", "efficiency" and "market economy"' (Coletta & Raftopoulos, 2016, p. 11). Similarly, new left governments, such as those in China, Brazil and Venezuela in the past decade, have endorsed concepts of growth in an attempt to compete with more established economies (Gudynas, 2011). The emergence of social indigenous groups and activist groups engaging in political spaces allows for the creation of new ecological narratives that revisit the concepts of modernity and assist in the decolonization of local knowledge and in the proposal of new ontologies.

Climate change also creates concern in the population. An increase in temperatures of 1.6–4°C is estimated by the end of the twenty-first century (Lachaud, Bravo-Ureta, & Ludena, 2022). For local communities living off the land, climate variability is one of the main perceptions of climate change, followed by temperature increase, changes in precipitation and changes in cultural habitats (Forero, Hernández, & Zafra, 2014). Changes in temperature will bring serious consequences for countries that are already experiencing socio-economic difficulties, including low economic growth, poverty, inequity, and debt (Cavallo, Hoffmann, & Noy, 2023). Despite this, Latin America has contributed to emphasizing the challenges of agriculture's role in natural resource management, the global impact of climate change on food systems and sustainable agricultural production (Coletta & Raftopoulos, 2016).

With all these ecological concerns in mind, this chapter seeks to shed light on the advances that the field of ecolinguistics has produced in Latin America. First, the term 'ecolinguistics' will be defined to establish a clear benchmark and ensure clarity in the terminology. Next, the hypotheses underpinning this chapter will be problematized in relation to the themes identified in the data. Finally, advances and challenges will be discussed in conversation with future lines of inquiry. The latter will hopefully serve as a springboard to inspire scholars to apply ecolinguistic approaches in their research.

Why (not) ecolinguistics?

Ecolinguistics as a subfield of linguistics has existed for over fifty years and can be argued to have its origins in the North American sphere with studies focusing mostly on language contact, minority languages and language loss, with Einar Haugen (1972) as the key figure. Arguably, Haugen's efforts focused mostly on the 'ecology of language' as a metaphor rather than the 'language of ecology', which distances itself from more contemporary approaches to the term 'ecolinguistics'.

One of the challenges of ecolinguistics is its undefined position within the field of studies. As Li, Steffensen and Huang (2020) point out, ecolinguistics has positioned itself within the scope of applied linguistics as an alternative to linguistics or a completely separate field of study (also noted in LeVasseur, 2015). With such opacity in terms of definition, Li, Steffensen and Huang (2020) set out to identify the research object and the research aims of the field in an attempt to offer clarity since it would otherwise be

> irrational to acknowledge all studies conducted under the label of the discipline first and then decide on the nature of the discipline based on the domain of these lines of research. By this reasoning, anything conducted under the label of the discipline becomes part of the discipline, leaving the discipline without focus or coherence. (p. 5)

In this sense, the underpinning principles behind ecolinguistics studies are the connections between language and ecology, both conceived as independent but interrelated fields. Nonetheless, the research methodology apparatus for ecolinguistic analysis remains limited. A survey conducted among members of the Language and Ecology Research Forum (now the International Ecolinguistics Association) found that a variety of methods are needed to conduct an ecolinguistic study (LeVasseur, 2015). Responses ranged from corpus approaches, conceptual metaphor theory, appraisal theory, framing and the reliance on other sub-disciplines, such as sociolinguistics, semiotics, pragmatics, functional linguistics and critical discourse analysis, as the most accepted methods. Just as linguistics nurtures many other areas of inquiry in the social sciences and the humanities (Derni, 2008), ecolinguistics also tackles a similar challenge, relying on knowledge and methods that extend the boundaries to both ecology and the hard sciences.

Since ecolinguistics can be a nuanced term, this chapter will primarily consider the definition set out by the International Ecolinguistics Association on its website:

> Ecolinguistics explores the role of language in the life-sustaining interactions of humans, other species and the physical environment. The first aim is to develop linguistic theories which see humans not only as part of society, but also as part of the larger ecosystems that life depends on. The second aim is to show how linguistics can be used to address key ecological issues, from climate change and biodiversity loss to environmental justice.[1]

This definition is useful for the purposes of this analysis since it is broad enough to encompass a variety of linguistic and discursive examples found in the data. Nevertheless, since the aim is to describe the main trends in contemporary research, this approach will not be prescriptive and will value the nuances of the term 'ecolinguistics' that have led to the development of ecolinguistic studies in Latin America. Furthermore, reflections will be given on whether there should be one or more ways of doing ecolinguistics and whether several approaches can coexist, particularly considering the disharmony that exists in European ecolinguistics (Zhou, 2017).

Given that the climate crisis needs local responses, it is also important to be critical of the ecolinguistic voice that Latin America is developing. As Mühlhäusler (2018) warns, developing a distinct vision of what doing ecolinguistics involves should take precedence over replicating models conceived overseas. An example of this is China, where harmonious discourse analysis (HDA) emerged as an alternative to other Western approaches to ecolinguistics and resonates within the Chinese context (Huang & Zhao, 2021). This is because HDA draws on Chinese philosophy to consolidate a more nuanced and tailored approach to ecolinguistics studies in that region.

Latin America has been part of the global conversation on sustainability since its inception (Vanhulst, 2019). While the influence of the Limits to Growth and the Brundtland Report remain iconic references, Latin American intellectuals have nonetheless been able to develop their own strands with a shared focus on transformation that conceives the social and environmental crises as rooted in the structures of modern society, often arising from a colonial and Eurocentric past. Therefore, discourses on post-development (Escobar, 1995) and alternatives such as Buen Vivir (Gudynas, 2011) emerge as a result of the process of reinterpretation of intellectual and political relationships from a local perspective, thus revitalizing the independence of marginalized cultural models and policies of development (Vanhulst, 2019).

But even knowledge construction is perceived differently across cultures. In discussing environmental justice decolonization, Álvarez and Coolsaet (2020) state that 'Non-Western communities are attached to the "empirical" or the material, while Western societies are able to provide the theoretical framework to conceptualize such practices' (p. 63). Therefore, recognizing that other types of knowledge outside of academic ones exist is an important step to encourage active participation from the Global South in the creation of knowledge. Despite looming concerns about climate change and the various socio-economic issues deeply attached to colonialism and natural resources extractivism, linguistic studies focusing on the ecological aspects arising from these issues remain marginal in the region (Zhang, 2022).

Methods

The underpinning hypothesis is that ecolinguistics has produced very limited impact in terms of research outputs in Latin America and in terms of establishing itself as a formal field of study independent from other discursive approaches (Forte, 2023). To test this, a review of the literature available online that incorporates the notion of 'ecolinguistics' in relation to Latin America was conducted. A search was performed in different databases in English, Spanish and Portuguese, the latter two being the most spoken languages in the region. The databases included Redalyc, Latindex, SciELO and Dialnet. The following keywords were used (in the three languages) to retrieve results: 'ecolinguistics', 'ecolinguistics in Latin America', 'ecological linguistics' and 'advances in ecolinguistics'. Since databases only provide a partial pool of available works (Siddaway, Wood & Hedges, 2019), the search was further refined with grey literature databases such as Google Scholar and on academic platforms like Academia.edu and ResearchGate to account for unpublished works.

The inclusion criteria comprised: (1) works that addressed the region and/or a specific area or country along with an explicit mention of ecolinguistics. However, since it was also plausible that there were outputs that fit the ecolinguistic criteria but that had not been labelled accordingly, a closer inspection was performed to identify specific cases of (2) outputs that used an ecolinguistics approach but did not explicitly include the term 'ecolinguistics'. This narrowed down the scope of the search even further.

The final sample consisted of a total of sixty-three outputs in Spanish (*N* = 35), Portuguese (*N* = 22) and English (*N* = 6). Spanish, the most spoken language in the region with around 200 million speakers, appears to be the most predominant language for publication, followed by Portuguese with just over 180 million speakers located mostly in Brazil. One of the main reasons for the interest in ecolinguistics in Brazil is due to the works of scholars like Hildo Honório do Couto, who has advocated for ecolinguistics since the late 1990s. He has extensively published in Portuguese, setting up the Brasilia School of Ecolinguistics in collaboration with the Universidade Federal de Goiás, a group of researchers from different areas of Brazil and other parts of the world that subscribe to ecosystemic linguistics and ecosystemic discourse analysis approaches. The website of the Grupo de Estudos e Pesquisas em Linguística Ecossistêmica condenses a comprehensive list of works published in Portuguese, including blogs, PhD and MA dissertations, reviews and news about events.[2] These efforts also translated into the creation of *Ecolinguística: Revista Brasileira de Ecologia e Linguagem Eco-Rebel*, a journal that since 2015 has regularly published works in the field and showcases a range of articles predominantly focusing on contact languages and developments in ecosystemic linguistics.

It is assumed that since the time the data was collected, the pool of available studies may be much broader and is therefore not accounted for. Representative examples of the sampled data are included throughout the chapter.

Results/findings

Advances in ecolinguistics in Latin America: A review

A total of seven themes were identified across the data: (1) ecolinguistics as an additional perspective; (2) ecolinguistics as language ecology; (3) no explicit mention of ecolinguistics but similarities in the analysis; (4) reviews and overviews of ecolinguistic works/approaches; (5) ecolinguistic analyses following a critical approach; (7) ecolinguistic analyses within the ecosystemic approach; and (7) minority languages and other practical analyses. The following sections outline the main findings in each category with some representative examples.

Ecolinguistics as an additional perspective

Theoretical frameworks set the basis for building knowledge in any research study and provide the structure and key concepts that align with the topic and the research questions underpinning the study (Grant & Osanloo, 2014). Therefore, introducing ecolinguistics within the theories that underpin a specific study is a relevant and much-needed exercise when it comes to deriving knowledge from other ecological humanities subjects. For instance, Singer and Villari (2016) study the interdisciplinary dialogue that is needed between linguistics and social geography to discuss topics relating to power, nature, society and education in their work with bilingual children in Peru speaking Quechua and Spanish. Ecolinguistics is introduced as a means to raise concerns about the loss of language and the loss of local knowledge embedded in Quechua, which often cannot be simply translated into another language.

In the context of anthropology, Skewes, Guerra and Henríquez (2014) draw on Mühlhäusler and Peace (2006) to embed ecolinguistics in their study as one of the many contributions to the understanding of human and environmental relationships. Pérez (2016) examines the legal framework of indigenous languages in Bolivia, Panama and Paraguay and draws on the works of European scholars (cf. Moure, 2012; García, 2011; Junyent, 2012) to emphasize that language loss implies losing versions of the world. To discuss globalization and interculturality within contemporary Spanish in Latin America, Ruiz (2010) introduces Calvet's (1999) gravitational model. These examples suggest a need to include linguistic frameworks that shed light on various ecological issues while illustrating how humanities and social sciences studies can benefit from incorporating ecolinguistic perspectives to deepen the theoretical analyses of their works, even if only as a secondary perspective.

Ecolinguistics as language ecology

Aside from the oft-cited lack of clarity in its research aims and object of study, ecolinguistics faces other challenges regarding the use of the prefix 'eco'. Stibbe (2021) notes that the term 'ecolinguistics' has been extensively used over the years:

> to describe studies of language interaction and diversity; studies of text such as signposts which are outdoor analysis of texts which happen to be about the environment; studies of how words in language relate to objects in the local environment; studies of the mix of languages surrounding pupils in

> multicultural schools; studies of dialects in particular geographical locations; and many other diverse areas. The multiplicity of approaches arises from different understandings of the concept of 'ecology', from a very broad concept of 'the interaction of some things with other things' to narrow concepts such as 'related to environmentalism'. (p. 8)

While the 'eco' of ecolinguistics may very well remain problematic, particularly as the field continues to expand, perhaps examining impact is one possibility to achieve some consistency. Therefore, 'Research on "language ecology" only becomes ecolinguistics when the impact of language contact on human behaviour and hence on real ecosystems is considered' (Alexander & Stibbe, 2014, p. 109). With such a wide range of understandings as to what counts and doesn't count as ecolinguistics over the years, it is not surprising that this fuzziness also applies to works in the Latin American context. In this sense, Couto and Couto (2016) approach ecolinguistics from a different angle:

> We believe that a better definition for the discipline is that Ecolinguistics is the study of the interactions between language and its social, mental, and natural context, via the people. Another way of saying the same thing would be: Ecolinguistics is the study of linguistic interrelations that take place at the social, mental, and natural levels. Both definitions avoid a reification of language, which is implicit in Haugen's definition. (p. 382)

One key representative example of this in the data is Stala (2017), who uses the concepts of language planning and status cultivation from Haugen (1972) to refer to the evolution of Argentinian Spanish.

No explicit mention of ecolinguistics but similarities in the analysis

As Krohn (2017) notes 'doing research in the context of real-world problems demands and develops types of skills and competencies that scholars are not used to' (p. 33). This often involves bringing different perspectives into the conversation, particularly when it comes to topics related to environmentalism. Some examples in the data reflect the similarities of the proposed analyses with that of ecolinguistics and could very well be marginally labelled as ecolinguistic analyses. Fill (1996) points out that 'some ecolinguists start at the ecology end and transfer ecological principles to language, while others start at the language end and bring linguistics to ecology' (cited in Couto, 2014, p. 123). This means that 'language' and 'ecology' are conceived as two independent but interrelated domains, 'investigating the relationship between language within the humanities and the discipline of ecology within the natural sciences' (Li, Steffensen, & Huang, 2020 p. 6). Such an approach would therefore suggest quite a broad research object for the field of ecolinguistics, an affordance which can nonetheless still be problematic.

In the surveyed data, Giraldo (2012) examines the use of metaphors such as 'nature as a resource' and 'the Earth as a mother' derived from the Pachamama concept in Latin America. While there is no explicit mention or reference to following an ecolinguistics approach, Giraldo's statement that 'language is a mediator between ourselves and the world we live in and that we share with others in society' (p. 223) can be understood as a broad approximation to Couto's (2016) triad of social, mental and natural contexts. In applying critical discourse analysis to the General Wildlife Law in Mexico, Salazar and García (2021) pursue a semi-ecolinguistic approach that would broadly align with ecological discourse analysis approaches. Similarly, this would suggest that more studies may remain unlabelled, which calls for more qualitative and interpretive investigations of what it entails to do ecolinguistics.

Reviews and overviews of ecolinguistic works/approaches

As with the development of any field, publications become available in different languages. In the case of the Latin American context, the accessibility of academic publications in ecolinguistics, particularly in Spanish – both in print and in electronic form – remain low. Therefore, reviews become a way to familiarize scholarly communities with the latest developments in a particular field. In the case of Portuguese, Costa (2015) presents a review of an interview with Hildo do Couto. Similarly, Sousa (2015) reviews Elza K.K.N do Couto et al.'s *Antropologia do imaginário, ecolinguística e metáfora*. Timbane (2018) reviews *Lingüística ecossistêmica: 10 anos de ecolingüística no Brasil* (Couto, Dourado, Silva & Avelar Filho, 2017). Overviews in Spanish focus on the contribution of ecolinguistics in the context of the Anthropocene (Farías & Reyes, 2020) and reviews of publications that are still unavailable in Spanish at the moment of writing (Roccia, 2020; ernández, 2023).

Ecolinguistic analyses following a critical perspective

Ecological discourse analysis (EDA) approaches have provided significant advances in the understanding of how discourse can enact both harmful and helpful messages, and can influence our relationship with the environment (Fill

& Penz, 2022). Within EDA, two main strands can be found: (1) the analysis of ecological discourse, or how we talk about ecology in the text-critical sense as proposed by Fill and Mühlhäusler (2001); and (2) the ecological analysis of discourse, which aligns with the system-critical strand advanced by Halliday (2001). Working with an ecological framework in mind – or an ecosophy in Naess's terms (1995) – is an essential element for ecolinguistics not to be merely considered a subset of discourse analysis (Alexander & Stibbe, 2014). Any discourse has the potential for conveying a worldview that encourages preservation or destruction of the ecosystems that sustain life; thus, focusing only on texts that explicitly focus on ecological concerns is limiting, since human behaviour is influenced by a multitude of underlying narratives, some of which are not as explicit in the discourse but which nonetheless have a powerful impact on the way we think and act. As such, working with an explicit ecological framework in mind will provide an otherwise lacking normative orientation when it comes to doing ecolinguistics.

In the Latin American context, the works of Diego Forte take many of these aspects on board. As the local representative of the International Ecolinguistics Association in Argentina and founder of the research group Ecolingüística Argentina, Forte's efforts have been pivotal in disseminating the field in Spanish in Latin America. Forte's works are influenced by critical discourse analysis with a focus on speciesism, semiotics and multimodality. Some key examples include the analysis of the audiovisual discursive construction of non-human animals in mass media by means of erasure and prominence (Forte, 2018), and tracing the story of meat in the construction of Argentina's national identity through the use of frames and multimodality (Forte, 2019a).

Other examples include theorizing about the role of ecolinguistics in deconstructing hegemonic roles. Most notably, Forte emphasizes that speciesist discourses are no different from other types of discourses that oppress human minorities. The notion of deconstructing damaging discourses and finding alternatives in non-European sources is a recurrent theme in the school of Latin American environmental thought (Leff, 2012). In this sense, Forte (2020) warns that 'the optimistic vision of Ecolinguistics remains to some extent as a romantic colonialist one because it conceives the other as the source of a potential different kind of development' (p. 100). This is a provocative point for enabling the kind of discussions that can further expand the field of EDA without perpetuating colonialist views, an aspect further elaborated in a more recent article in relation to the weak formal establishment of ecolinguistics in higher education institutions in Latin America (Forte, 2023).

But even the exercise of shifting the narrative with non-European alternatives or deconstructing existing ones will remain problematic if the Marxist notion of class isn't reformulated, an aspect that ecolinguistics has to take on board. The edited volume *Lenguaje and Biodiversidad* (Forte, 2019b) features a range of contributions that may fall within the borders of ecolinguistics since 'Questioning those social representations that promote the destruction of other species and their surroundings is sufficient to be considered within the boundaries of Ecolinguistics' (Forte, 2019b, p. 16). Therefore, reformulating old notions that perpetuate colonialist and hegemonic views that oppress and destroy both the environment and other species will be within the purview of what counts as ecolinguistics, an aspect that closely aligns with Alexander and Stibbe's (2014) conceptualization of the ecological analysis of discourse. *Lenguaje y Biodiversidad* presents works that cover the representation of human and non-human species (see Franco San Román, 2019; Murguía Cruz, 2019; and Perla, 2019) and environmental representation, with distinct contributions focusing on the identity of Costa Rica (cf. Hernández Sánchez, Hinkelammert Palma and Sánchez Sánchez 2019) and discourses about mining and tourism in Costa Rica (cf. Rivera Alfaro, 2019).

Ecolinguistic analyses within the ecosystemic approach

Ecosystemic linguistics appears to be the most widespread approach in Latin America, particularly thanks to the efforts of the Brasilia School of Ecolinguistics. A large portion of the surveyed data explicitly draws on Couto's theory of ecosystemic linguistics (Couto, 2007) and ecological discourse analysis with key theoretical developments (Couto & Couto, 2016; Couto, 2020; da Fonseca Fernandes, 2020), as well as applications from different angles, including ethnoterminology analyses (Costa & Gomes, 2013); historical linguistics and missionary linguistics (Teixeira & Kaltner, 2021); applications in the field of first-language acquisition (Couto & Fonseca Fernandes, 2013); ecolinguistics and the national curricular common base for Portuguese language teaching (da Silva & Couto, 2020); migration and language contact in Portuguese and Venezuelan Spanish (Silva & Guisan, 2019; Guisan & da Silva, 2022); and grammatical studies (Couto, 2014, 2016).

Other research has focused on the role of memory as a key feature for the existence of the linguistic ecosystem (Couto & Couto, 2020) and applying a linguistic ecosystem approach to rural language in Portuguese (Couto, 2018), including the social and natural environment of gypsies in Aparecida de Goiânia

(Couto, 2013). There are also studies that are based on a semantic analysis of the use of proverbs in legal texts at the intersection of ecosystemic linguistics and legal ecology (Andrade, 2020) with a specific interest in '*legal ecolinguistics*'. Along a similar line, Freitas Brandão, Belaidi and Couto (2020) compare speeches from countries with a biocentric approach and an anthropocentric approach towards nature. The findings suggest that the personification of nature leads to a 'distorted biocentrism' that covertly reproduces anthropocentric and capitalist ideals.

The number of Spanish sources following Couto's ecosystemic approach was lower than Portuguese sources. Some representative examples include studies on the use of language in Andean communities concerning discourses around biodiversity (Jiménez, 2020), indigenous Aymara children (Gonzalez Estay, 2021) and translanguage in Spanish and Quechua on Facebook and WhatsApp (Pérez, 2020). Other examples explore the development of an ecolinguistic conceptual model based on General Systems Theory (Estay Gonzalez, 2018), as well as analyses of the Brasilia School of Ecolinguistics (Ballesteros, 2015) and cosmoteandrism perspectives in ecolinguistics (Zunino, 2013).

Minority languages and other practical analyses

Latin America and the Caribbean home approximately 600 indigenous languages,[3] many of which face the threat of extinction due to globalization and homogenizing processes (Pérez, 2016). Therefore, research has focused on linguistic diversity, language contact and linguistic policies, particularly in the use of Spanish, Portuguese and English over local minority languages (Zimmermann, 2011). This is how ecolinguistics is framed in some cases, as an approach, a theoretical framework, that sustains the contention that the loss of linguistic diversity will entail the loss of local knowledge embedded in that language. This area features a more consistent pattern of interest in applying ecolinguistic frameworks due to the tangible need to protect those affected languages and their communities.

Studies have focused on the use of English as a lingua franca threatening minority languages, as seen in Despagne Broxner and Sánchez Martínez's (2021) analysis of linguistic ideologies at a Mexican research centre: adopting Skutnabb-Kangas' (2011) multilingual approach allows the ecological perspective to reveal the tension and pressures of using Spanish, English and other languages in academia. On the other hand, Ardila, Angel and Loaiza-Camacho (2021) examine the threat posed by the advance of Spanish to the Kriol language in San

Andrés, Providencia and Santa Catarina, in Colombia. Bondarenko Pisemskaya (2010a, 2010b) identifies several contributors and processes that increase linguistic diversity of Venezuela and emphasizes the need to avoid cultural and linguistic homogenization.

Drawing from Fill and Mühlhäusler (2001), Tapia Kwiceien, Gelvez Anello and Solla (n.d.) apply ecolinguistics to analyse the language policies for teaching Spanish as a mother tongue or as a foreign language within the Argentinian educational system, and they propose didactic exercises to incorporate ecolinguistics to 'allow students to reflect on linguistic diversity and multilingualism and, in doing so, raise awareness of environmental issues' (p. 5). But the same holds for Spanish as a dominant language: Stork (2008) incorporates ecolinguistic concepts in reflecting on the importance of linguistic diversity when comparing Hispanic Spanish lexicography in Latin America with European counterparts.

Discussion

Challenges and opportunities for ecolinguistics in Latin America

Just as ecolinguistics is a heterogeneous field in the Anglo world, these similarities also appear to translate to Latin American production only on a micro scale. Advances in ecolinguistics in Latin America are necessary, and changes are taking place, though at a lower rate than expected, considering the rapid advances of the climate crisis. As with many other emerging fields, such as environmental humanities, there can be a lag in terms of 'converting' people into the field. The lack of formal establishment that the discipline has had on the continent, with a few exceptions, can be accounted as one of the reasons for this (Forte, 2023).

Questioning why there hasn't been more active engagement across the region would be unfair given the vastness and often fragmented nature of the continent, and the well-organized achievements of the Brasilia School of Ecolinguistics, for instance, and the local production of individual scholars. The sampled data covers ecolinguistic research from countries such as Brazil, Argentina, Chile, Ecuador, Bolivia, Peru, Costa Rica, Mexico, Colombia and Venezuela, which confirms that ecolinguistics is being pursued across the continent. At the time of writing, a new regional centre of the International Ecolinguistics Association has been established in Uruguay, the third regional centre based in South America alongside Brazil and Argentina, and another promising step in the development of the field in the region.

It has perhaps been the expectation that, given the multitude of realities, ecolinguistics would have been engaged in more evenly across the continent. But attempts can be seen in the sampled documents where scholars incorporate ecolinguistics into the analysis of key local concerns, predominantly language contact and language loss, and from disciplines such as anthropology (Skewes, Guerra, & Henríquez, 2014) and social geography (Singer & Villari, 2016).

Similarly, the lack of agreement as to what is or is not ecolinguistics, while not often reflected explicitly in the outputs, also suggests, in some cases, a reliance on the advances proposed by European perspectives. The problem lies in the often-cited lack of uniformity or consensus in the field (Zhou, 2017), particularly in Western traditions. For ecolinguistics to be a truly representative discipline, it must be made up of scholarly contributions from all over the world, including in non-European languages (Penz & Fill, 2022). This also involves taking a stance in the debate as to what is and isn't ecolinguistics, and developing a local voice that resonates with the needs of specific parts of the world. In this sense, ecosystemic linguistics appears to be the leading trend in the region with an interesting and varied range of applications in fields such as ethnoterminology, migration and language contact, historical linguistics and legal (eco)linguistics, illustrating the ductility of this approach.

But would it be fair to assert that there should be one model of ecolinguistics that can be applied across the continent? While there are many issues shared by Latin American countries alike, the reality of each country varies considerably. As a rather novel discipline, attempting any kind of prescriptive approach will not bode well in areas of the world that have long suffered from the pains of colonialism and extractivism, both in the linguistic and physical realms. The climate crises affect us differently, and just as solutions must come from collaboration across disciplines, ecolinguistics in Latin America will have a different flavour from that in other parts of the world.

A degree of scepticism can also arise, particularly from the lack of engagement from 'linguistic proper' or applied linguistics, and from continuing to apply theories conceived in the European tradition. The exercise of merely transferring Western concepts and frameworks to other regions can be counterproductive and do more harm than good, particularly in the case of environmental justice movements (Álvarez & Coolsaet, 2020) and arguably other fields. As Huang and Zhao (2021) note in the context of Chinese ecolinguistics:

> With increasing attention given to ecological problems worldwide, different ideas and schools from different regions will bring vitality to ecolinguistics, at the same time as leading to misunderstandings. Therefore, it is important and necessary for us to draw out the universal implications from the philosophy of a certain culture. We believe this is an effective way to promote dialogue between West and East. (p. 16)

If the concept of environmental thinking is still being developed in Latin America as a means of gaining identity in the face of the environmental crisis, then it is no surprise that ecolinguistics is also finding its feet in a more equally distributed manner in Latin America. Further explorations in the field of ecological critical analysis could shed more light into decolonization and work with an indigenous second-sight that could also transform reality (Rappaport & Pacho, 2012) – also raised by Corbetta (2021). As Ballesteros (2015) points out, the risk of Couto's approach, if any, lies in neutralizing language from any linguistic-political positioning from the linguist (p. 136). In this sense, Ballesteros insists that bringing ecolinguistics into conversation with critical linguistics points us in the direction of linguistic policies and the focus on the norm. While implicit, ecolinguistics also has a normative approach towards protecting the flourishing of life (Stibbe, 2021). Therefore, keeping a balance between the descriptive and more critical aspects is crucial for a more uniform development of the field.

Conclusion

This chapter sought to trace the current trends in research on ecolinguistics in or about Latin America. It started with the premise that ecolinguistic research in the field has been produced marginally, and therefore, it sought to trace evidence available online of works that can reflect the state of the art. The data surveyed in Spanish, Portuguese and English broadly demonstrated seven different areas in which research has predominantly focused or in which the term 'ecolinguistics' has been applied: (1) ecolinguistics as an additional perspective; (2) ecolinguistics as language ecology; (3) no explicit mention of ecolinguistics but similarities in the analysis; (4) reviews and overviews of ecolinguistic works/approaches; (5) ecolinguistic analyses following a critical approach; (6) ecolinguistic analyses within the ecosystemic approach; (7) minority languages and other practical analyses.

It is unfair to state that ecolinguistics has not emerged in Latin America, since that would undermine the efforts of those working tirelessly to collate knowledge and encourage participation on a regular basis. The opportunities lie in the further development of investigations into what can work for the field in regional languages. The interest in developing a Latin American perspective in terms of environmental concerns is not a new subject; therefore, integrating linguistics, particularly ecolinguistics, into the conversation is a crucial aspect for giving more linguistic precision to studies that incorporate it from outside of linguistics. Further opportunities lie in the development of theories and ontologies that align with socio-economic and ecological local concerns, and that can assist in framing so-called Latin American environmental thought. In turn, this would enable interdisciplinary collaboration networks that can strengthen the field of ecolinguistics at multiple levels of participation, thus gaining more presence in the region.

Notes

1 https://www.ecolinguistics-association.org/
2 ecoling.unb.br
3 https://www.worldbank.org/en/region/lac/brief/indigenous-latin-america-in-the-twenty-first-century-brief-report-page

References

Alexander, Richard, & Arran Stibbe. (2014). From the analysis of ecological discourse to the ecological analysis of language. *Language Sciences*, *41*, 104–110.

Álvarez, L., & Coolsaet, B. (2020). Decolonizing environmental justice studies: A Latin American perspective. *Capitalism Nature Socialism*, *31*(2), 50–69. https://doi.org/10.1080/10455752.2018.1558272

Andrade, T. L. S. (2020). Onde háfumaça, háfogo: os provérbios no Texto Jurídico entre a sabedoria popular, a Ecolinguística e o Direito. *Travessias*, *14*(1), 170–184.

Ardila, R. S., Angel, M. S., & Loaiza-Camacho, B. S. (2021). Retos y oportunidades contemporáneos del kriol en el archipiélago de San Andrés, Providencia y Santa Catalina, Colombia. *Forma y Función*, *34(2)*. https://doi.org/10.15446/fyf.v34n2.88613

Ballesteros, A. (2015). La Ecolingüística en Brasil. Teoría del Lenguaje y Descripción de la Lengua: Aportes para una política del portugués brasileño. *Revista Digital de Políticas Lingüísticas (RDPL)*, *7*, 127–138.

Bondarenko Pisemskaya, N. (2010a). Situación ecolingüística venezolana contemporánea. Papeles de trabajo-Centro de Estudios Interdisciplinarios en Etnolingüística y Antropología. *Socio-Cultural*, (20), 22–35.

Bondarenko Pisemskaya, N. (2010b). Lenguas minoritarias de Venezuela: Consideraciones desde la perspectiva ecolingüística. *Revista de Filología y Lingüística de la Universidad de Costa Rica, 36*(1), 175–189.

Calvet, L. (1999). *Per una écologie des langues du monde*. Plon.

Cavallo, E., Hoffmann, B., & Noy, I. (2023). Disasters and climate change in Latin America and the Caribbean: An introduction to the special issue. *Economics of Disasters and Climate Change, 7*(2), 135–145. https://doi.org/10.1007/s41885-023-00132-2

Coletta, M., & Raftopoulos, M. (Eds.). (2016). *Provincializing nature: Multidisciplinary approaches to the politics of the environment in Latin America*. Institute of Latin American Studies. Retrieved from September 22, 2023, from https://www-jstor-org.glos.idm.oclc.org/stable/10.2307/j.ctv13nb6g6

Corbetta, S. (2021). Educación Ambiental y Educación Intercultural: hacia una construcción de puentes desde un pensamiento ambiental y latinoamericano crítico. *Gestión y ambiente, 24*(supl1), 107–130.

Costa, N. M. (2015). Ecolinguística: Um diálogo com Hildo Honório do Couto-2013. *Ecolinguística: Revista brasileira de ecologia e linguagem (ECO-REBEL), 1*(1), 145–146.

Costa, N. M. P., & Gomes, D. M. (2013). A etno terminologia da Língua Mundurukú-Tupí e as contribuições da Ecolinguística. *Cadernos de Linguagem e Sociedade, 14*(1), 252–274.

Couto, E. K. N. N. do. (2018). Abordagem linguístico-ecossistêmica da linguagem *rural: uma primeira aproximação. Revista de Letras, Fortaleza, 37*(2), 79–88.

Couto, E. K. N. N. do, & Couto, H. H do. (2016). Ecolinguística, lingüística ecossistêmica e análise do discurso ecológica (ADE). *Signótica, 28*(2), 381–404.

Couto, E. K. N. N. do, Dourado, Z., Silva, A. N. da, & Avelar Filho, J. N. (Eds.). (2017). Linguística ecossistêmica: 10 anos de ecolinguística no Brasil. Pontes.

Couto, E. K. N. N. do., & Couto, H. H. do. (2020). A memória e a existência do ecosistema linguístico. *Travessias, 14*(1): 185–197. https://doi.org/10.48075/rt.v14i1.24114

Couto, E. K. N. N. do, & da Fonseca Fernandes, E. M. (2013). Aquisição de língua: uma perspectiva ecolinguística. *Letras de Hoje, 48*(3), 290–298.

Couto, H. H. do. (2007). *Ecolinguistica: estudo das relações entre língua e meio ambiente*. Thesaurus.

Couto, H. H. do. (2014). Ecological approaches in linguistics: A historical overview. *Language Sciences, 41*, 122–128.

Couto, H. H. do. (2016). Estudos gramaticais à luz da lingüística ecossitêmica. *Scripta, 20*(38), 26–53.

Couto, H. H do. (2020). Análise do discurso ecossistêmica–ADE. *Árboles y Rizomas, 2*(2), 1–14.

da Fonseca Fernandes, E. M. (2020). Concepções de discurso na Análise do Discurso Ecossistêmica. *Travessias, 14*(1), 235–245.

da Silva, E. P., & do Couto, E. K. N. N. (2020). Entrelaçamentos: ecolinguística e Base Nacional Comum Curricular para o Ensino de Língua portuguesa. *Travessias, 14*(1), 140–153.

Derni, A. (2008). The ecolinguistic paradigm: An integrationist trend in language study. *The International Journal of Language Society and Culture, 24*, 21–30.

Despagne Broxner, C. I., & Sánchez Martínez, E. (2021). Ideologías lingüísticas en un centro de investigación en México. *Sinéctica*, (56), 2–21.

Escobar, A. (1995). *Encountering development: The making and unmaking of the third world*. Princeton University Press.

Estay, C. R. G. (2018). Modelo conceptual ecolingüístico basado en la teoría general de sistemas (TGS): una propuesta. *Ecolinguística: Revista brasileira de ecologia e linguagem (ECO-REBEL), 4*(1), 59–75.

Estay, C. R. G. (2021). *Niños indígenas Aymaras de la Región de Tarapaca* (Doctoral dissertation, Universidad de Valladolid).

Farías, M., & Reyes, S. (2020). Estudios lingüísticos y literarios en el marco del Antropoceno. *Arbolesy Rizomas, 2*(2), i–vi.

Fill, A. (Ed.). (1996). *Sprachökologie und Ökolinguistik.*Tübingen.

Fill, A. & Mühlhäusler, P. (Eds.). (2001). *The ecolinguistics reader: Language, ecology and environment*. Continuum.

Fill, Alwin F., & Penz, H. (2017). Ecolinguistics in the 21st century: New orientations and future directions. In *The Routledge handbook of ecolinguistics* (pp. 437–443). Routledge.

Forero, E. L., Hernández, Y. T., & Zafra, C. A. (2014). Percepción latinoamericana de cambio climático: metodologías, herramientas y estrategias de adaptación en comunidades locales. una revisión. *Revista U.d.c.a Actualidad & Divulgación Científica, 17*(1), 73–85.

Forte, D. L. (2018). De la naturaleza a su mesa: borramiento, naturalización y deconstrucción en el conflicto cresta roja, *Revista de Letras, 37*, (2), 162–178.

Forte, D. L. (2019a). La construcción de discurso identitario nacional argentino: la historia de la carne. *Question / Cuestión, 1*(64), 25.

Forte, D. L. (Ed.). (2019b). *Lenguaje y Biodiversidad: Nuevas perspectivas para viejas desigualdades*. Ecolinguistica Argentina.

Forte, D. L. (2020). Ecolingüística y la nueva lucha de clases: contra la especie dominante. *Pensamiento al margen: Revista Digital de Ideas Políticas, 12*(11), 90–102.

Forte, D. L. (2023). Latin American ecolinguistics: Deconstructing discourse studies, coloniality and industrial environmentalism. *Journal of World Languages, 10*(2), 330–349. https://doi.org/10.1515/jwl-2023-0052.

Franco San Román, M. (2019). De Lassie a Cujo: La metáfora del perro como dispositivo de la construcción del êthos de Cleón en la obra aristofánica. In D. L. Forte (Ed.), *Lenguaje y biodiversidad: Nuevas perspectivas para viejas desigualdades* (pp. 39–68). Ecolingüística Argentina.

Freitas Brandão, H. de, Belaidi, R., & Couto, E. K. N. N do. (2020). A (in) efetividade jurídica da personificação da natureza: uma análise ecolinguística das medidas jurídicas de proteção ambiental. *Travessias, 14*(1), 246–265.

García, A. (2011). Ecoloxía e lingua: unha estraña parella? *Moenia, 17*, 491–518.

Giraldo, O. F. (2012). El discurso moderno frente al "pacha ma mismo": La metáfora de la naturaleza como recurso y el de la Tierra como madre. *Polis. Revista Latino americana, 11*(33), 219–234.

Grant, C., & Osanloo, A. (2014). Understanding, selecting, and integrating a theoretical framework in dissertation research: Creating the blueprint for your "house". *Administrative Issues Journal, 4*(2), 4.

Gudynas, E. (2011). Buen vivir: today's tomorrow. *Development, 54*(4), 441–447.

Guisan, P. G. F., & da Silva, M. I. A. (2022). Migraçã ona fronteira Brasil-Venezuela: Interações e uso de marcadores discursivos. *Ecolinguística: Revista brasileira de ecologia e linguagem (ECO-REBEL), 8*(2), 126–142.

Haarstad, H. (Ed.). (2012). *New political spaces in Latin American natural resource governance*. Springer.

Halliday, Michael A. K. ([1990] 2001). New ways of meaning: The challenge to applied linguistics. In Alwin Fill & Peter Mühlhäusler (Eds.), *The ecolinguistics reader: Language, ecology and environment* (pp. 175–202). Continuum.

Haugen, E. (1972). *The Ecology of Language. Language science and national development.* Stanford University Press.

Hernández, C. D. (2023). Reseña. Ecolinguistics: Language, ecology and the stories we live by. *Arboles y Rizomas, 5*(1), 101–104.

Hernández Sánchez, M., Hinkelammert Palma, V., & Sánchez Sánchez, D. (2019). La representación de la identidad costarricense en la marca país esencial Costa Rica: Una mirada desde la teoría del realce en textos multimodales. In D. L. Forte (Ed.), Lenguaje y biodiversidad: Nuevas perspectivas para viejas desigualdades (pp. 101–122). Ecolingüística Argentina.

Huang, G., & Zhao, R. (2021). Harmonious discourse analysis: Approaching peoples' problems in a Chinese context. *Language Sciences, 85*, 1–19.

Jiménez, M. A. (2020). Dinámicas territoriales y discursos sobre biodiversidad en una comunidad andina de Bolívia. *Ecolinguística: Revista brasileira de ecologia e linguagem (ECO-REBEL), 6*(1), 52–82.

Junyent, M. C. (2012). Para sermos lingüisticamente sustentábeis. In Sanmartín, G. (Ed.), *VIII Xornadas sobre Lingua e Usos* (pp. 129–140). Universidade da Coruña.

Krohn, W. (2017). Interdisciplinary cases and disciplinary knowledge. In R. Frodeman, J. T. Klein, & R. C. D. S. Pacheco (Eds.), *The Oxford handbook of interdisciplinarity* (pp. 32–49). Oxford University Press.

Lachaud, M. A., Bravo-Ureta, B. E., & Ludena, C. E. (2022). Economic effects of climate change on agricultural production and productivity in Latin America and the Caribbean (lac). *Agricultural Economics, 53*(2), 321–332. https://doi.org/10.1111/agec.12682

Leff, E. (2012). Latin American environmental thinking: A heritage of knowledge for sustainability. *Environmental Ethics, 34*(4), 431–450.

LeVasseur, T. (2015). Defining "Ecolinguistics?": Challenging emic issues in an evolving environmental discipline. *Journal of Environmental Studies and Sciences, 5*, 21–28.

Li, J., Steffensen, S. V., & Huang, G. (2020). Rethinking ecolinguistics from a distributed language perspective. *Language Sciences, 80*, 101277. https://doi.org/10.1016/j.langsci.2020.101277

Moure, T. (2012). Do activismo ecolóxico ao activismo lingüístico. In Sanmartín, G. (Ed.), *VIII Xornadas sobre Lingua e Usos* (pp. 65–84). Universidade da Coruña.

Mühlhäusler, P. (2018). Peter Mühlhäusler. *Ecolinguística: Revista brasileira de ecologia e linguagem (ECO-REBEL), 4(*1), 140–143.

Mühlhäusler, P. y Peace, A. (2006). Environmental discourses. *Annual Review of Anthropology, 35*, 457–479.

Murguía Cruz, N. (2019). "Somos la voz de los que no tienen voz": Representaciones dominantes y alternativas sobre los animales. In D. L. Forte (Ed.), Lenguaje y biodiversidad: Nuevas perspectivas para viejas desigualdades (pp. 69–82). Ecolingüística Argentina.

Naess, A. (1995). The shallow and the long range, deep ecology movement. In A. Drengson & Y. Inoue (Eds.), *The deep ecology movement: an introductory anthology* (pp. 3–10). North Atlantic Books.

Penz, H., & Fill, A. (2022). Ecolinguistics: History, today, and tomorrow. *Journal of World Languages, 8*(2), 232–253.

Pérez, I. C. (2016). Lenguas indígenas, identidad y marco legal en Hispanoamérica: una aproximación ecolingüística. *E-Aesla*, (2), 88–200.

Pérez, V. L. (2020). El translenguaje digital, estrategia discursiva ecológica de jóvenes bilingües quechua" "castellano en Facebook y WhatsApp. *Ecolinguística: Revista brasileira de ecologia e linguagem (ECO-REBEL), 6*(1), 83–103.

Perla, L. (2019). ¿Cómo se relaciona la opresión que sufrimos lxs sujetxs no hegemónicos (machismo) y la explotación animal (especismo)? In D. L. Forte (Ed.), Lenguaje y biodiversidad: Nuevas perspectivas para viejas desigualdades (pp. 83–100). Ecolingüística Argentina.

Rappaport, J., & Pacho, A. R. (2012). Collaboration and historical writing: Challenges for the indigenous-academic dialogue. In F. Mallon (Ed.), *Decolonizing native histories: Collaboration, knowledge, and language in the Americas* (pp. 122–143). Duke University Press.

Rivera Alfaro, L. (2019). De discursos y realidades: Minería, turismo, producción piñera y resistencias en Costa Rica. In D. L. Forte (Ed.), Lenguaje y biodiversidad: Nuevas perspectivas para viejas desigualdades (pp. 143–160). Ecolingüística Argentina.

Roccia, M. (2020). Reseña. The Routledge Handbook of Ecolinguistics. Introducción de Alwin F. Fill, edición de Alwin F. Fill y Hermine Penz. *Árboles y Rizomas, 2*(1), 96–98.

Ruiz, C. S. (2010). Español actual: Globalización e interculturalidad. *Decires, 12*(14), 75–89.

Salazar, J. I. C., & García, B. A. B. (2021). La fauna silvestre en los discursos de la Ley General de Vida Silvestre, su reglamento y de los inspectores ambientales de México. *Nóesis. Revista de Ciencias Sociales, 30*(60), 104–125.

Siddaway, A. P., Wood, A. M., & Hedges, L. V. (2019). How to do a systematic review: A best practice guide for conducting and reporting narrative reviews, meta-analyses, and meta-syntheses. *Annual Review of Psychology, 70,* 747–770. https://doi.org/10.1146/annurev-psych-010418-102803

Silva, M. I. A., & Guisan, P. F. (2019). Dos contatos iniciais ao bom da migração venezuelana: aspectos linguísticos. *Revista Internacional de apoyo a la inclusión, logopedia, sociedad y multiculturalidad,* 5(3), 120–128.

Singer, K., & Villari, C. (2016). Uniendo burbujas de pensamiento. Búsqueda de diálogo interdisciplinario entre una lingüista y una geógrafa sobre el trabajo con niños en una región de que chuah ablantes. *INDIANA-Estudios Antropológicos sobre América Latina y el Caribe, 33*(1), 69–90.

Skewes, J. C., Guerra, D., & Henríquez, C. (2014). Patrimonio y paisaje: dos formas de ensamblar naturaleza y cultura en la cuenca del río Valdivia, sur de Chile. *Chungará (Arica), 46*(4), 651–668.

Skutnabb-Kangas, T. (2011). Language ecology. In J. O. Östman y J. Verschueren (Eds.), *Pragmatics in practice.* John Benjamins Publishing Company.

Sousa, R. M. (2015). Antropologia do imaginário, ecolinguística e metáfora-2014. *Ecolinguística: Revista brasileira de ecologia e linguagem (ECO-REBEL), 1*(1), 149–151.

Stala, E. (2017). El idioma de los argentinos: entre Borges y Bordelois. *Verba Hispanica, 25*(1), 133–146.

Stibbe, A. (2021). *Ecolinguistics: Language, ecology and the stories we live by* (2nd ed.). Routledge.

Stork, Y. (2008). La situación lexicográfica del español hispanoamericano. Consideraciones desde la perspectiva de la ecolingüística. In Azorín Fernández, D. et al., *El diccionario como puente entre las lenguas y culturas del mundo. Actas del II Congreso Internacional de Lexicografía Hispánica, Alicante* (pp. 594–600). Biblioteca Virtual Miguel de Cervantes.

Tapia, K., Gelvez, A., & Solla. (n.d). *Ecolingüística para el estudio y enseñanza del español lengua materna y lengua extranjera.* Retrieved from: https://rdu.unc.edu.ar/bitstream/handle/11086/1780/Tapia%20Kwiecien%20M.%20Gelvez%20Anello%20V.%20Solla%20J.%20Ecolinguistica%20para%20el....pdf

Teixeira, V. L., & Kaltner, L. F. (2021). Comunidades de fala efêmeras: uma leitura ecolinguística da carta de Caminha. *Cadernos de Linguística, 2*(4), e504–e504.

Timbane, A. A. (2018). Dez anos de Ecolinguística no Brasil. *Ecolinguística: Revista brasileira de ecologia e linguagem (ECO-REBEL), 4*(1), 128–131.

Vanhulst, J. (2019). Pensar la sustentabilidad desde América Latina. Retrospectiva del discurso académico a partir de un análisis bibliométrico entre 1970 y 2012. *Revista Colombiana de Sociología, 42*(1), 41–71. https://doi.org/10.15446/rcs.v42n1.73141

Zhang, R. (2022). The year's work in ecolinguistics 2021. *Journal of World Languages*, 8(1), 141–163. https://doi.org/10.1515/jwl-2022-0009

Zhou, W. (2017). Ecolinguistics: Towards a new harmony. *Language Sciences, 62*, 124–138.

Zimmermann, K. (2011). La construcción ecolingüística del contacto de lenguas (español y lenguas amerindias). In Yolanda Congosto Martín & Elena Méndez García de Paredes (Eds.), *Variación lingüística y contacto de lenguas en el mundo hispánico* (pp. 361–388). Vervuert Verlagsgesellschaft.

Zunino, F. (2013). Una Visión Holística de la Naturaleza cultura: El Cosmote andrismo en los Glifos de las Ciudades Prehispánicas de Habla Nahuatl. *Cadernos de Linguagem e Sociedade, 14*(1), 149–174.

2

Investigating the discrepancies between environmental education and students' environmental practices

A mixed-method approach

Hassane Razkane, El Ayachi El Baghdady,
Adil Youssef Sayeh and Mohamed Yeou

Introduction

Environmental education plays a fundamental role in the promotion of sustainable practices and the shaping of students' behaviours towards their surroundings (Novotný et al., 2021). It is regarded as a tool that can resolve the growing environmental problems by introducing effective strategies to reduce issues regarding this topic (Estrada-Araoz et al., 2023). Environmental activists, policymakers and researchers all call for the enhancement of individuals' awareness and sensitivity. Understanding the discrepancies between environmental education and students' undesirable behaviours, on the one hand, and determining the root factors behind these unfriendly practices, on the other, can be beneficial for future environmental interventions (Balundė et al., 2019). Also, measuring students' environmental behaviours in schools is critical to comprehending how educational initiatives can improve students' environmental practices and potentially drive societies towards sustainability (Mónus, 2022). Thus, exploring the relationship among education, environmental awareness and students' behaviours, and identifying the complex interplay of influential factors that leads to misbehaviours (Ahmat Zainuri et al., 2022) demand a thorough investigation in order to come up with practical and adequate solutions.

The Moroccan context emerges as a compelling area of research to explore the discrepancies between environmental education and students' unfriendly environmental behaviours. Aligning with the global community's calls for sustainable practices, it is imperative to understand the nature and specificities of the gap between environmental education in Morocco and students' everyday ecological practices. This will provide further insights into how to raise students' awareness to improve their daily practices towards their surroundings. The examination of the multifaceted factors influencing Moroccan students' environmental behaviour, such as littering, excessive energy use, water wastage, improper disposal of waste, disregard for green spaces and lack of participation in environmental initiatives, may highlight the barriers constraining effective practices and assess the effectiveness of current educational practices.

To unravel these disparities, determine key factors behind such actions and unearth some practical solutions to improve students' behaviours, the current chapter aims to delve into this nexus by focusing on three primary objectives. First, it seeks to identify and assess the undesirable environmental practices prevalent among Moroccan middle-school students; second, it aims to evaluate students' awareness of the environmental practices they are involved in; and finally, it endeavours to explore the factors contributing to the observed discrepancies between education and students' actual actions with the aim of unearthing practical solutions to improve student behaviours.

The particular research described in this paper adopts a social perspective on Moroccan society, inquiring into psychological influences shaped by cultural norms and individual perceptions that play a pivotal role in determining students' interactions with the environment. It also investigates the institutional dynamics within educational settings that contribute to the shaping of environmentally conscious behaviours. By dissecting these factors, we aim to provide a comprehensive understanding of the intricate web influencing the environmental consciousness of Moroccan students.

This chapter aims to provide valuable insight into the field of environmental education and sustainability, offering practical recommendations for educators, policymakers and stakeholders to foster more sustainable practices among middle-school students and inform tailored interventions and policies to promote sustainable practices in the Moroccan context. Thus, this chapter intends to pave the way for sustainable practices suited to the unique socio-cultural and environmental landscape of the country. Based on these objectives, three main research questions guide the current work:

1. What are middle-school students' undesirable environmental practices?
2. To what extent are these students aware of their environmental practices?

3. What are the underlying factors behind the discrepancy between environmental education and students' actions towards the environment?

Literature review

Socio-ecological systems theory

Understanding the complex interactions between human behaviour and the environment is crucial for sustainable development and environmental conservation. In this sense, this chapter adopts a socio-ecological approach to describe the nuances of middle-school students' environmental behaviours as well as the motives behind their unfriendly practices towards it. This theoretical framework offers a thorough lens through which to address the abovementioned research questions, emphasizing the relationship between environmental education, personal actions and the larger socio-ecological setting. The paradigm offered by socio-ecological systems theory (SEST) (Berkes et al., 2003; Folke et al., 2010; Ostrom, 2009; Ostrom and Cox, 2010) allows for the investigation of the complex interplay of institutional, societal and personal elements impacting behaviour.

The SEST looks at the dynamic interactions between social and ecological elements within a particular system. It investigates the interdependence of social, psychological and institutional factors, recognizing the influence of cultural norms, community dynamics and institutional factors on students' environmental practices (Berkes and Folke, 1998; Ostrom, 2009). SEST investigates the intricate relationships that exist between social and ecological systems. It acknowledges that human cultures and the ecosystems in which they live are inextricably linked and impact each other in dynamic ways, triggering changes in particular components that can have a domino effect on the entire system. This theory highlights the importance of studying these interactions as integrated systems rather than as isolated entities. It examines the ways in which institutions, values and human activity impact and are impacted by the environment, emphasizing the necessity of an integrated and flexible approach to dealing with complex environmental problems. SEST acknowledges that in order to foster long-term stability, well-being and effective sustainable resource management, it is significant to gain a deeper understanding of both social and ecological dynamics influencing individuals' behaviours (Berkes and Folke, 1998; Ostrom, 2009).

Focusing on the adaptability and resilience of socio-ecological systems, SEST examines how systems involving both people and the environment can adapt and bounce back from challenges, particularly because these systems deal with interruptions such as changes in the environment or social unrest. SEST also delves into the role of institutions and governance structures in shaping how humans interact with the environment to manage resources and achieve a certain degree of sustainability, asserting that different governance systems influence this management and, therefore, the degree of sustainability achieved or desired. SEST takes into account feedback loops that can either provide stability or exacerbate changes. It also discusses the idea of cross-scale dynamics, examining the way in which systems operate at different scales, from local to global, and how changes at one level can impact other levels. SEST incorporates the concept of panarchy, which defines adaptive cycles of change at various socio-ecological system scales. Additionally, understanding the reciprocal influence human behaviour has over the ecosystem is a crucial component of human-environment interaction. For more details see Ostrom (2009).

In summary, SEST aligns well with the multifaceted nature of the current study, offering a lens to examine how social influences (e.g. peer interactions, family norms), psychological factors (individual beliefs) and institutional dynamics (school policies, curriculum) collectively shape Moroccan students' environmental behaviours.

Students' environmental awareness and practices

Considerable research has delved into students' awareness of environmental protection, disclosing divergent results. While some findings underscored insufficient knowledge among middle-school students (Choe et al., 2020) as well as low awareness in senior high school students (Ata, 2018), others highlighted moderate knowledge and awareness (Abilo and Linaugo, 2018; Gök and Firat Kiliç, 2021) to high levels of awareness among middle-school students (Baig, 2020; Bhowmik, 2019). These findings indicate that middle-school students' environmental awareness ranged from low and moderate to high, which requires further research to account for these discrepancies.

On the other hand, a study investigating the environmental sensitivity and awareness of university students (Oguz et al., 2010) indicated that despite undergoing multiple courses on environmental issues, students' awareness and responsible behaviours were below the expected levels. The conclusion drawn by researchers was that possessing environmental knowledge, including the understanding of concepts and facts, did not consistently translate into heightened awareness nor evidenced a deeper perception or consciousness regarding environmental issues and behavioural intentions reflecting individuals' planned actions or conducts involving environmental matters. Consistent findings were reported by Abbas and Singh (2014), who indicated that despite holding positive attitudes, students exhibited low involvement in environmental activities. Other studies uncovered that although secondary school students exhibited a generally high level of environmental awareness, there were discrepancies between environmental education and students' involvement in pro-environmental activities (Altin et al., 2014). Although the majority of undergraduate students had a strong understanding of environmental issues, only a few actually engaged in pro-environmental behaviours (Handoyo et al., 2021). These results show the discrepancies between education and actual ecological behaviours and imply that other factors could be at play.

In conclusion, the literature reviewed here provides a deeper vision of students' environmental awareness and practices. While some inconsistencies exist, the overall findings underscore the importance of fostering environmental awareness to promote positive behaviours towards sustainability. Further research is needed for a better description of these gaps and to enhance our understanding of the complex relationship between awareness and environmental practices among students.

Factors behind environmental practices

Some research has examined the factors behind students' ecological practices such as conserving resources, reducing waste and protecting ecosystems. Socio-economic backgrounds, including the socio-economic status of schools, residences and parents' education levels, emerged as key factors influencing students' behaviours towards the environment (Mónus, 2022). Further results supporting the link between environmental education and pro-environmental behaviours were reported by Estrada-Araoz et al. (2023), who pointed out a direct and significant relationship between education and pro-environmental behaviours among Peruvian students. Similarly, Ahmat Zainuri et al. (2022) suggested that educating children about conservation behaviours could play a crucial role in influencing their parents with low socio-economic backgrounds to adopt more environmentally friendly practices. Furthermore, Yusliza et al. (2020) found that environmental commitment, consciousness and green lifestyle positively influenced pro-environmental behaviour, emphasizing the need for enhancing students' knowledge and behaviours aligned with environment-focused education. In a textual analysis of fourteen environmental texts, Mliless and Larouz (2018) suggested that these textbooks may not effectively convey the true nature of environmental challenges, potentially hindering a comprehensive understanding of ecological issues among students.

Methods

Participants

The study detailed in this chapter focused on middle-school students in El Jadida province, Morocco, with a total of 122 participants (63 males and 59 females). All of the participants were in either in their 8th grade (forty-five participants) or 9th grade (seventy-seven participants) and were selected through stratified random sampling with representation from two different schools: one private (fifty-seven participants) and the other public (sixty-five participants). The participants' ages spanned the formative period between thirteen and sixteen ($M = 13.94$, $SD = .87$), belonging to different socio-economic statuses. Twenty participants from each grade were selected through stratified random selection for the interview. For ethical reasons, the identities of the respondents were concealed and referred to using pseudonyms.

Data collection instruments

Our investigation employed a robust data collection approach, utilizing a questionnaire administered to approximately 122 middle-school students. Additionally, semi-structured interviews were conducted with twenty informants to gather information on middle-school students' undesirable environmental practices and to get an insight into the underlying factors behind students' undesirable actions towards the environment, eliciting effective strategies needed to bridge the gap between students' knowledge and their daily behaviour. The questionnaire contained twelve items along with the demographic data. It unfolds in two major distinct sections, apart from the items devoted to the collection of the participants' demographic information. The first part has five items that explore students' environmental practices at school. It carefully probes into the realm of students' behaviours, providing a clear description of these practices. The second segment contains seven items aimed at determining the key factors behind students' behaviours. These items unravel the intricacies of the factors influencing behaviours, encapsulating the individual within the social, psychological and institutional dynamics. A five-point Likert scale, ranging from 'I strongly agree' to 'I strongly disagree' was used for rating. The questionnaire was administered in Arabic.

In line with the quantitative approach, a semi-structured interview was employed as a poignant tool, engaging twenty informants – ten males and ten females – adding a qualitative layer to our exploration. These in-depth interviews were designed for triangulation purposes, aligning with the questionnaire to enrich our understanding of the nuances within the data. The interview included eight major items with follow-up questions. Along with a background section which served to collect information regarding the respondents' names, genders, ages, educational levels, types of school, residence, educational levels of parents and socio-economic statuses of parents, the interview contained three parts. The first section namely had two central questions aimed at disclosing students' environmental behaviours. The second part of the interview sought to determine the key factors behind undesirable environmental practices. It included two key questions regarding social factors, two about psychological motives and two items about institutional factors. All the items were measured based on a five-point Likert scale.

Procedures

Prior to data collection, a meticulous design phase was undertaken to craft both the comprehensive questionnaire and a set of semi-structured interview questions. To check clarity and eligibility, a preliminary test was conducted with a sample group of students. The questionnaire and interview questions that raised concerns or were deemed unclear were refined for optimal comprehension. As the pilot study progressed, ongoing refinement of the questionnaire and interview approach was undertaken. Feedback from participants, as well as researchers' observations, contributed to iterative improvements in the data collection instruments, further enhancing the study's reliability and validity.

Upon confirmation of the clarity and relevance of the questionnaire, the researchers administered it to the target participants. Elaborated in Standard Arabic, the questionnaire was distributed in a public and a private school to ensure representativeness and include participants from various socio-economic backgrounds. The participants were allotted approximately half an hour to complete a paper-based questionnaire in Standard Arabic. This timeframe was aimed at striking a balance between allowing participants sufficient time to consider their responses and maintaining the research's efficiency. Clear instructions were provided, elucidating the purpose of the research and emphasizing the significance of accurate and thoughtful responses. To mitigate ambiguity, students were encouraged to seek clarification on any questions that posed difficulties during the completion process.

Similarly, the semi-structured interviews, lasting twenty minutes on average, were conducted with twenty informants and followed a rigorous process. The purpose of these interviews was to get a deeper understanding of students' environmental practices, the motives behind undesirable environmental behaviours and the strategies needed to promote pro-environmental behaviours. The interviews were conducted in Moroccan Arabic. The students were approached, informed about the research's objectives and given the option to participate. First, approval was granted by the school administration of the target schools, and then students were given the opportunity to take part in the study. Prior to initiating the interviews, explicit details and clarifications were offered, ensuring participants understood the nature of the study. The participants were informed about the approximate duration beforehand, promoting informed consent and cooperation. Throughout the interviews, participants were encouraged to seek explanations for any unclear queries, fostering an environment conducive to open and transparent communication.

Throughout both the questionnaire and interview phases, ethical considerations were paramount. Informed consent was obtained from all participants, and their right to withdraw from the study at any point was reiterated. Confidentiality measures were implemented to safeguard the participants' privacy and ensure the integrity of the collected data.

These meticulous data collection procedures were implemented to ensure the quality, clarity and ethical integrity of the research process, fostering an environment conducive to meaningful participant engagement and accurate representation of their perspectives.

Data analysis

The quantitative data obtained from the questionnaire was described via descriptive statistics, given this study is exploratory in nature. This approach allowed us not only to paint a vivid picture of students' environmental behaviours and the underlying factors behind their ecological practices but also to discern the relationships between these variables. As for the qualitative findings, the richness of the interview data was unfurled through thematic coding. Thematic analysis was applied to the twenty audio-recorded interviews, with two researchers ensuring accuracy through transcription, translation and coding. Themes and subthemes, informed by the questionnaire and the theoretical framework, were deduced to answer the research questions. The qualitative data provided diverse perspectives on students' environmental practices, the motives behind such behaviours and the strategies needed to minimize students' unfriendly environmental behaviours. This meticulous process involved developing codes to categorize and distil the essence of the narratives, offering a nuanced perspective on the qualitative dimensions of the study. Through the marriage of quantitative rigour and qualitative depth, this work endeavours not only to uncover the discrepancies in environmental education but also to illuminate pathways for fostering a culture of environmental protection within the Moroccan educational landscape.

Results

Middle-school students' environmental practices

The analysis of both the questionnaire responses and the interview findings revealed concerning tendencies in students' environmental practices within the school. More than two-thirds (70.5 per cent) of the respondents stated that students' behaviour has a negative impact on the cleanliness of the school premises. Instances of disruptive behaviours, such as writing on walls, throwing garbage in corridors, leaving water taps open, turning on the lights in the corridor and rooms during daylight, and sabotaging and breaking desks, are highlighted in the interviews.

Excerpt 1

> Students don't respect the school premises. They're mostly writing on the wall, throwing garbage in corridors, squares and sections??, leaving water taps open, turning on lights in daylight, and sabotaging and breaking desks. (Respondent 1)

Furthermore, environmental practices that are essential for promoting sustainability within the school environment, including activities such as reducing waste, recycling, conserving water and energy, and minimizing pollution, were also examined. Results showed that approximately 63.9 per cent of participants expressed the belief that students frequently disrespect school property and resources, and a significant 70.5 per cent of respondents indicated that students deliberately violate school rules related to environmental protection. Additionally, a striking 87 out of 122 participants admitted to witnessing instances of unnecessary paper waste and littering, indicating a prevalent issue that needs addressing.

The interviews provide specific examples of this disrespect, including writing on walls and desks, trampling on plants, picking plants, damaging trash bins, breaking desks, tearing and throwing papers under tables, littering handouts in classrooms, yards and corridors, and damaging green spaces in both the schoolyard and nearby parks.

Excerpt 2

> I've seen students trampling on plants, picking plants and damaging the trash. They do graffiti and break desks. Most of them leave the taps running and leave waste everywhere. (Respondent 2)

Excerpt 3

> When students eat something, they throw their handouts anywhere, whether they're in the classroom, the yard or the corridors. (Respondent 12)

Students' awareness of their actions towards the environment

The analysis of responses to questions regarding individual actions and awareness of their environmental impact provides a deeper understanding of the perceptions and consciousness of participants towards the environment. When asked about the extent to which their individual actions contribute to the overall environment, a good number of respondents (58.2 per cent) acknowledge being aware of their impact on the environment. However, a notable 39.5 per cent admit to not being aware of the significant consequences of their actions on the environment, while 12.3 per cent remain neutral in their stance. This diversity in responses suggests varying levels of environmental consciousness among the participants.

Out of the eleven interview questions provided, four questions were about individual respondents' behaviour (questions 3, 5, 8 and 9), while seven questions were related to the respondents' observations of other students' behaviours (questions 1, 2, 4, 6, 7, 10 and 11). From the interview data, some participants articulate a heightened awareness, noting the regular exposure to environmental issues in the media. This articulates a higher level of information absorption through their exposure to media coverage of environmental issues. Some of the participants in the interview further elaborate on the adverse outcomes such as air pollution, acid rain, unpleasant odours and pollution-related diseases resulting from practices that harm the environment, and give examples emphasizing how the disappearance of a forest affects air quality in some given areas.

Excerpt 4

> Every day we watch on television and in the news the effects of environmental pollution, high temperatures, floods, lack of precipitation, drought (Respondent 14)

Excerpt 5

> Air pollution, acid rain, the spread of unpleasant odours and the spread of pollution-related diseases are the results of our practices that are harmful to the environment. (Respondent 11)

These examples vividly detail some respondents' understanding of various negative outcomes linked to harmful environmental practices, demonstrating a deeper understanding of the repercussions their actions have on the welfare of their surrounding environment. They add an ecological perspective, emphasizing the interconnectedness of ecosystems and the impact of actions on air quality.

On the other hand, only a few express a lack of awareness about the detrimental effects of their actions on the environment. They openly admit to feeling insufficiently aware of the dangers of their environmentally harmful actions and emphasize the general lack of awareness in the community.

Excerpt 6

> There isn't enough awareness of the consequences of harmful actions to the environment. (Respondent 14)

Different factors influencing students' negative behaviours towards the environment

Peer influence

The data revealed divergent results concerning the influence of peers on individuals' actions and practices towards the environment. The quantitative data analysis indicates that 34.4 per cent of respondents admit being influenced by their peers to engage in environmentally undesirable practices, while 51.6 per cent assert that their peers have no influence on their environmental practices. This divergence in responses suggests a varied range of susceptibility to peer influence among the participants.

The interview data unveiled that some respondents acknowledge feeling influenced by their colleagues, indicating that peer pressure sometimes leads them to engage in practices harmful to the environment. One of them shares an incident where a friend's suggestion led him to throw a piece of paper onto the street because there already was visible garbage around.

In short, both the quantitative and qualitative findings clearly indicate the influence of peers' unfriendly ecological behaviours on an individual's daily practices at school and in the street.

Excerpt 7

Lack of facilities to protect the environment, lack of trash cans and garbage containers. Also, lack of interest in green spaces and afforestation. (Respondent14)

Excerpt 8

There is a lack of concern for the environment. The school does not punish violators. I think the absence of strictness in dealing with bad practices is harmful to the environment. And this encourages pupils to do harmful things to the environment. (Respondent 7)

School influence

The exploration of students' opinions on the effectiveness of school policies in encouraging environmental respect reveals a mixed perspective. About a third (32 per cent) of respondents recognize that their schools apply some rules to deter undesirable environmental practices, while a similar percentage (35.1 per cent) declares the opposite. Additionally, 23 per cent are neutral on the effectiveness of school policies. This indicates a diversity of opinions regarding the impact of school rules on students' behaviour towards the environment.

The interview data indicate that participants cite various concerns related to the perceived inadequacy of school policies. There is a consistent theme highlighting a lack of concern for the environment within the school, coupled with the observation that the school does not adequately punish violators or implement deterrent penalties. This perceived leniency in dealing with practices harmful to the environment is considered a deterrent to encouraging positive behaviour.

The current research also probed into whether environmental school facilities such as trash cans and garbage containers are adequate. In a related question about the sufficiency of such facilities to safeguard the environment, 43.4 per cent of respondents assert that their schools contain the necessary facilities, while 38.6 per cent state the opposite and 18 per cent remain neutral.

The interviews raise concerns about the perceived lack of facilities for environmental protection, specifically noting insufficient trash cans and garbage containers. There is also a noted lack of interest in green spaces and afforestation, suggesting that the physical aspects of the school environment may not be adequately equipped to promote and sustain environmentally friendly practices.

In summary, data analysis of both the questionnaire and the interviews suggests that there are various factors influencing students' negative behaviours towards the environment, exposing distinct trends in peer, family, community and school influences. Peer influence varies among participants, with some acknowledging the impact of classmates on engaging in environmentally undesirable practices. Family dynamics play a significant role, with participants noting that parental attitudes towards waste management shape their behaviours. Community attitudes strongly affect students' environmental practices at school, with concerns raised about the lack of environmental awareness and a general disregard for the environment within society. School policies and facilities also contribute to shaping behaviour, with mixed perspectives on the effectiveness of rules and concerns about inadequate facilities for environmental protection. These trends highlight the interconnected nature of social, familial and institutional influences on students' environmental behaviours, emphasizing the importance of addressing these factors comprehensively for fostering positive environmental practices.

Excerpt 9

I see the influence of where we live, the city is full of pollution and there is a lack of interest in waste management. (Respondent14)

Excerpt 10

Society doesn't care about the environment and this is reflected in the behaviour of students towards the environment at school. (Respondent 17)

Excerpt 11

We notice waste everywhere, there is no awareness of the environment, and this affects the behaviour of students and their environmental behaviour within the institution. (Respondent 18)

Community influence

The investigation of the factors contributing to the consistency or difference in students' environmental practices at school reveals a predominant influence of community attitudes on their behaviours. From the questionnaire data, it emerges that 61.5 per cent of respondents declare that their community attitudes towards the environment significantly influence their environmental practices at school. In contrast, only 15.6 per cent state that the community has no impact on their environmental practices, with 23 per cent remaining neutral. This may indicate a strong link between community attitudes and students' environmental practices at school.

Several factors contribute to this observed influence of the community. The participants expressed concerns about the lack of environmental awareness and a general disregard for the environment within the community. Different participants emphasize that practices within the school are greatly influenced by societal norms and note a significant lack of concern for the environment, with students showing indifference to designated waste disposal areas. They highlight the pervasive presence of waste, indicating a lack of awareness and concern for the environment within society. These social practices are echoed at school, and societal disregard for the environment is reflected in students' behaviour at school.

Additionally, the informants identify a lack of awareness, ignorance of the seriousness of their actions and an absence of deterrent sanctions as contributing factors. Zakaria emphasizes that the city is dominated by pollution, and there is little interest in waste management, reinforcing the notion that community practices significantly impact students' environmental behaviour.

To conclude, the findings from both the questionnaire and the interviews highlight the critical influence society exerts on a student's environmental behaviours. The unfriendly ecological attitudes and daily practices prevalent in the community are found to have an influential impact on individuals' perceptions of and practices towards the environment.

Excerpt 12

This is related to upbringing, if the family is keen to raise its children well, this is reflected in their behaviour. (Respondent 10)

Excerpt 13

In the house, students make sure that it is clean, the house is private, while the school is a public space. At home, I feel censored, while at school there is the freedom to act and do anything because there is no supervision. (Respondent 14)

Excerpt 14

Students don't deal in the same way, at home there is respect and appreciation for the environment because of fear, punishment, blame or reprimand, while at school there is a large margin of freedom that makes students indifferent to their actions towards the environment. (Respondent 11)

Excerpt 15

My family does not show conscious behaviour towards the environment and this affects my behaviour. (Respondent 7)

Excerpt 16

When my father is driving his car, he opens the car window and throws his waste in the street. (Respondent 11)

Excerpt 17

My parents promote pro-environmental practices, when I leave waste in an inappropriate place, they tell me to put it in the waste bin. (Respondent 18)

Excerpt 18

As a young child, you are influenced by your parents, and when you notice your family's interest in waste management, this is reflected in your practices. (Respondent 10)

Family influence

The investigation into the influence of family practices on students' environmental behaviour at school provides significant insights into the role of familial dynamics in shaping attitudes towards the environment. When asked about the impact of their family's environmental practices at home on their behaviour at school, 54.9 per cent of respondents acknowledged this influence. Only 29.5 per cent perceived little to no impact from their families on their environmental practices at school.

The qualitative data analysis provides more insights into how families impact students' behaviour. Some respondents emphasize the formative role of parental influence, noting that a child's practices are shaped by observing their family's interest in waste management. This underscores the pivotal role of family attitudes and practices in moulding environmentally conscious behaviours in students. Others also note that their family actively cares about the environment, guiding them when they improperly dispose of waste and engage in waste collection and burning, indicating an effort in managing waste.

However, other cohorts highlight a nuanced situation where there is a relative interest in environmental practices within the narrow confines of the home but express concerns about family members discarding waste inappropriately outside, reflecting a limited awareness of broader environmental responsibility. Some of them even lament the lack of conscious behaviour towards the environment within their family.

Regarding whether students exhibit similar environmental behaviours at school as they do at home, the respondents expressed a consensus that there is a difference in behaviour between home and school environments. One participant observes that fear, punishment, blame or reprimand in the home setting encourages respect and appreciation for the environment, whereas the school environment, in which students perceive themselves to have a larger margin of freedom, may make students indifferent to their actions. Another informant notes the contrast between the private nature of home and the public space of the school, where a perceived lack of supervision allows for more freedom in actions. Also, one respondent attributes this difference to upbringing, stating that a family's commitment to raising children well is reflected in their behaviour.

In summary, the results from the questionnaire and the interview revealed that parental behaviours towards the environment can have a tremendous impact on children's everyday ecological practices.

Excerpt 19

One time I was walking with my classmates and I had a piece of paper in my hand, I was waiting until I found the garbage bin, and one of my friends told me to dispose of it, the street is full of garbage, so I threw it away. (Respondent 10)

These findings highlight a spectrum of awareness among respondents regarding the environmental impact of individual actions. While some participants evidence a heightened consciousness, others express a lack of awareness or neutrality. The qualitative responses provide additional context, revealing varying degrees of understanding regarding the consequences of environmentally harmful practices. These practices not only contribute to environmental degradation but also pose a challenge to maintaining a clean and organized learning environment. They not only signify a lack of regard for the school's physical infrastructure but also contribute to an environment that is less conducive to learning and overall well-being. This underscores the tangible consequences of these actions on the school premises and highlights a systemic issue requiring attention to reinforce the importance of adhering to established rules for the benefit of the entire school community.

Discussion

This chapter analyses students' environmental behaviours and awareness, and the diverse factors that impact their environmental practices within the school setting. Organized thematically, the results provide a thorough exploration of these crucial aspects, addressing the three primary research questions separately.

Consistent with previous studies (Abilo & Linaugo, 2018; Baig, 2020; Bhowmik, 2019; Gök & Firat Kiliç, 2021), our results reveal that despite expressing a heightened consciousness about the impact of their actions on the environment (Choe et al., 2020), the majority of students engage in harmful practices. These practices have adverse effects on the overall cleanliness of the school, contributing to an environment that is less conducive to learning and overall well-being. We can conclude that awareness of good environmental practices does not necessarily translate into adopting positive behaviours (Abbas & Singh, 2014; Handoyo et al., 2021; Oguz et al., 2010). While students are well aware of the importance of the environment in their lives and the community as a whole, they lack environmental commitment and do not adopt an environmentally friendly lifestyle (Yusliza et al., 2020). This underscores the importance of targeted environmental education initiatives to enhance awareness and foster a more informed and responsible approach to individual actions.

Our findings also highlight that peers have a lesser degree of influence on students' behaviours compared to the effect exerted by the surrounding community and family. While peers exhibit divergent influences on students' environmental practices, family members are found to have a more significant role in moulding environmentally conscious behaviours in students (Ahmat Zainuri et al., 2022).

However, the surrounding community's significant lack of concern for the environment instils a sense of indifference in students towards the environment. The lack of environmental awareness, indifference to waste management and the absence of deterrent measures within the community contribute to the observed behaviours within the school environment (Yusliza et al., 2020). Addressing these community-level factors is crucial for fostering a culture of environmental responsibility among students and promoting positive environmental practices within the school setting. These social practices are echoed at school, emphasizing the effect of societal disregard for the environment on students' behaviour at school.

Concerning the influence of school policy and educational programs, the data points to different concerns regarding their efficacy and adequacy (Mliless & Larouz, 2018). They are found to deter students from engaging in positive behaviours towards the environment. The perceived lack of stringency in enforcing rules and the inadequacy of facilities to support environmental protection (Altin *et al.*, 2014) indicate areas for improvement within the school system. Addressing these concerns is crucial for creating an environment that not only educates but actively encourages positive environmental practices among students (Estrada-Araoz et al., 2023).

Hence, the variety of responses revealed by the data reinforces the need for targeted interventions to promote positive peer influence and cultivate a collective sense of responsibility for the environment within the student community. There is a need for comprehensive environmental education programmes that address varying family attitudes and behaviours, emphasizing the importance of instilling a sense of responsibility and commitment. These programmes should also include a comprehensive environmental education that extends beyond school to individual households. These programmes could include family environmental education workshops, home sustainability challenges, family eco-action plans, community-based environmental projects, educational resources for families, environmental education events and festivals, and school-home partnerships. There should be a connection between practices at home and school. In addition to that, the negative impact of community attitudes on students' engagement in good environmental practices at school implies that these programmes should also target community-level factors to foster a culture of environmental responsibility among students and promote positive environmental practices within the school setting.

Conclusion

The findings of the study in this chapter point to the necessity of comprehensive methods for effective environmental education. Mere dissemination of knowledge is insufficient. There is an urgent need for pedagogical strategies that go beyond delivering information to actively promote a sense of responsibility, agency and empathy towards the environment. To cultivate a pro-environmental culture, teachers should support students not only in learning about the importance of environmental protection but also in encouraging them to translate their environmental knowledge into tangible and sustainable practices. The current research also invites policymakers and educational institutions to devise effective educational initiatives and provide students with practical experiences that go beyond the school setting, helping them become responsible citizens towards the environment.

The findings suggest several significant implications for future research in the field of environmental education. First, researchers may examine and evaluate the design and effectiveness of current pedagogical strategies aimed at cultivating a sense of responsibility, agency and empathy towards the environment among students. Additionally, longitudinal studies tracking the long-term impact of environmental education programmes beyond middle school could provide insights into the sustainability of educational interventions. Investigating the role of peer influence and cultural norms in shaping environmental behaviours, and understanding how these dynamics evolve over time, presents another promising avenue for research. Moreover, fostering cross-disciplinary collaborations between environmental educators, psychologists and sociologists could offer holistic perspectives and innovative solutions to bridge the gap between environmental education and actual behavioural change. By addressing these research directions, scholars can contribute to the ongoing refinement and enhancement of environmental education initiatives for more effective and lasting impact.

References

Abbas, M. Y., & Singh, R. (2014). A survey of environmental awareness, attitude, and participation amongst university students: A case study. *International Journal of Science and Research (IJSR)*, *3*(5), 1755–7060.

Abilo, J. M. E., & Linaugo, J. D. (2018). Environmental knowledge and sensitivity of senior high school students. *Journal of Science Teachers and Educators*, *2*(1).

Ahmat Zainuri, N., Abd-Rahman, N., Halim, L., Chan, M. Y., & Mohd Bazari, N. N. (2022). Measuring po-environmental behavior triggered by environmental values. *International Journal of Environmental Research and Public Health, 19*(23). https://doi.org/10.3390/ijerph192316013

Altin, A., Tecer, S. Tecer, L., Altin, S., & Kahraman, B. F. (2014). Environmental awareness level of secondary school students: A Sase Study in Balıkesir (Türkiye). *Procedia – Social and Behavioral Sciences, 141*, 1208–1214. https://doi.org/10.1016/j.sbspro.2014.05.207

Ata, E. (2018). Evaluation of adult environmental awareness behaviours in terms of social learning theory according to perceptions of primary and secondary school students. *International Journal of Higher Education, 7*(6), 54–62. https://doi.org/10.5430/ijhe.v7n6p54

Baig, M. M. (2020). A study of environmental awareness ability of X Std. students of Urdu medium school of Aurangabad city. *International Journal of Reflective Research in Social Sciences, 3*(1), 14–16.

Balundė, A., Perlaviciute, G., & Steg, L. (2019). The relationship between people's environmental considerations and pro-environmental behavior in Lithuania. *Frontiers in Psychology, 10*, 1–10. https://doi.org/10.3389/fpsyg.2019.02319

Berkes, F., & Folke, C. (1998). *Linking social and ecological systems: Management practices and social mechanisms for building resilience*. Cambridge University Press.

Berkes, F., Colding, J., & Folke, C. (2003). *Navigating social-ecological systems: Building resilience for complexity and change*. Cambridge University Press.

Bhowmik, A. (2019). Environmental awareness among senior secondary school students: A comparative study. *Education India Journal: A Quarterly Refereed Journal of Dialogues on Education, 8*(2), 54–67.

Choe, J. H., Kim, C. H., & Ri, G. H. (2020). An investigation on the environmental knowledge and attitudes of senior middle school students in the democratic people's republic of Korea. *International Research in Geographical and Environmental Education, 29*(2), 146–162. https://doi.org/10.1080/10382046.2019.1678276

Estrada-Araoz, E. G., Gallegos Ramos, N. A., Paredes Valverde, Y., Quispe Herrera, R., & Mori Bazán, J. (2023). Examining the relationship between environmental education and pro-environmental behavior in regular basic education students: A cross-sectional study. *Social Sciences, 12*(5). https://doi.org/10.3390/socsci12050307

Folke, C., Carpenter, S. R., Walker, B., Scheffer, M., Chapin, T., & Rockström, J. (2010). Resilience thinking: Integrating resilience, adaptability and transformability. *Ecology and Society, 15*(4), 20.

Gôk, N. D., & Firat Kiliç, H. (2021). Environmental awareness and sensitivity of nursing students. *Nurse Education Today, 101*. https://doi.org/10.1016/j.nedt.2021.104882

Handoyo, B., Astina, I. K., & Mkumbachi, R. L. (2021). Students' environmental awareness and pro-environmental behaviour: Preliminary study of geography students at state university of Malang. *IOP Conference Series: Earth and Environmental Science, 683*(1). https://doi.org/10.1088/1755-1315/683/1/012049

Mliless, M., & Larouz, M. (2018). An ecolinguistic analysis of environment texts in Moroccan English language teaching textbooks. *Ijres*, *5*, 103–116.

Mónus, F. (2022). Environmental education policy of schools and socioeconomic background affect environmental attitudes and pro-environmental behavior of secondary school students. *Environmental Education Research*, *28*(2), 169–196. https://doi.org/10.1080/13504622.2021.2023106

Novotný, R., Huttmanová, E., Valentiny, T., & Kalistová, A. (2021). Evaluation of environmental awareness of university students: The case of the university of Presov, Slovakia. *European Journal of Sustainable Development*, *10*(2), 59–72. https://doi.org/10.14207/ejsd.2021.v10n2p59

Oguz, D., Cakci, I., & Kavas, S. (2010). Environmental awareness of university students in Ankara, Turkey. *African Journal of Agricultural Research*, *5*(19), 2629–2636.

Ostrom, E. (2009). A general framework for analyzing sustainability of social-ecological systems. *Science*, *325*, 419–422.

Ostrom, E., & Cox, M. (2010). Moving beyond panaceas: A multi-tiered diagnostic approach for social-ecological analysis. *Environmental Conservation*, *37*(4), 451–463. https://doi.org/10.1017/S0376892910000834

Yusliza, M. Y. et al. (2020). An investigation of pro-environmental behaviour and sustainable development in Malaysia. *Sustainability (Switzerland)*, *12*(17), 1–21. https://doi.org/10.3390/su12177083

3

The integration of ecolinguistics in the EFL context in higher education in Oman to raise students' motivation and environmental awareness

Iman Jabbar Abbas and Ouarda Khouni

Introduction

UNESCO has called for plans and strategies to integrate Global Citizenship Education (GCE) into education curricula to address the values of environmental education (EE) and education for sustainability (ES) (Jacobs et al., 2016). This trend in education encourages educators, curriculum developers and teachers to play a significant role in raising student awareness regarding how to deal with ecological issues and challenges (Fill & Penz, 2018).

In the field of linguistics, ecolinguistics has emerged as a new paradigm that aims to explore the relationship between language, ecology and society (Zhou, 2022). Hence, ecolinguistics must be introduced to the curriculum to raise students' awareness of the need to preserve natural resources in the environment where they live and beyond. Ecolinguistics aims to help students become ecological thinkers to understand the concept of environmental education and how to deal with the environment. More specifically, ecolinguistics uses language as a strategy or tool that contributes to raising student awareness of ecology.

Within the domain of the EFL context, ecolinguistics can be practised as an approach to teaching English as a foreign language (EFL) (Luardini & Sujiyani, 2018). From an ecological perspective, students will learn English terms that are associated with the environment and natural resources. In this regard, Zahoor and Janjua (2020) reported that English-language teaching (ELT) should take into consideration both the enhancement of language skills on the one hand and the raising of students' knowledge and awareness of environmental education on the other hand. This suggests that it has become crucial to raise students' awareness about current environmental issues, such as global warming and other environmental issues, when teaching EFL.

Based on the researchers' teaching experience in EFL teaching and curriculum development, it could be mentioned that there is a gap or an obvious absence of the approach of ecolinguistics in the curriculum of EFL, at least in the context where the current chapter takes place. The major goal of EFL, in the context of the work, is to offer English for academic purposes (EAP) and aims to improve students' linguistic competency and English-language communication skills. Realistically, there is a lack of awareness among undergraduates on topics that are related to globalization, the environment, human resources, and the like. No doubt, this lack of awareness will have negative impacts on communication skills and linguistic competency. How can an individual engage in a debate or discussion if he or she doesn't have some information of certain topics? In essence, EFL pedagogy needs to be shaped and guided by the approach of ecology or ecolinguistics in order not to be separated from the outside world. The EFL curriculum must include topics on current global or environmental issues to keep students updated with national and international environmental issues.

This chapter acknowledges the importance of integrating environmental education into the existing curriculum of EFL at the foundation level at the University of Technology and Applied Sciences in Oman. The rationale for the integration is to help students develop their language skills on the one hand and to become aware of their environment on the other hand. Another rationale is to contribute to the objectives of UNESCO, which encourage educational institutes worldwide to consider sustainable or environmental development.

Going through the literature, the researchers realized that a plethora of studies in the field of EFL had been conducted to investigate and address a variety of issues related to improving English communication skills and competency. However, there is a lack of research that investigates strategies, methods and techniques for integrating the concept of ecology into the EFL curriculum in the context of Oman. Specifically, the chapter aims to:

1. Investigate the pedagogical impacts of integrating the approach of ecolinguistics into the EFL curriculum from the perspectives of students in the EFL higher education context.
2. Explore the strategies and methods of integrating ecolinguistics into the curriculum of EFL to raise students' environmental awareness.

The chapter is guided by the following research questions:

1. How does the proposed integration of ecolinguistics affect students' motivation and engagement in learning from the perspectives of students in the UTAS foundation programme?
2. What strategies and methods must be implemented when integrating ecolinguistics into the EFL curriculum from the perspectives of students?

Literature review

Ecolinguistics refers to the role of language in studying ecological contexts that are concerned with the interactions of humans, other species and the physical environment for sustaining life (Stibbe, 2021). This emphasizes the contribution of language in the development of the surrounding environment and the provision of solutions for ecological problems. A more comprehensive description of ecolinguistics is provided by Stibbe (2021), stating that:

> The term 'ecolinguistics' has been used to describe studies of language interaction and diversity; studies of texts such as signposts which are outdoors; analysis of texts which happen to be about the environment; studies of how words in a language relate to objects in the local environment; studies of the mix of languages surrounding pupils in multicultural schools; studies of dialects in particular geographical locations, and many other diverse areas.
>
> (p. 8)

In the area of linguistics and sociolinguistics, ecolinguistics is considered a new paradigm that goes beyond associating language with only social contexts. Rather, it aims to associate language with wider ecological contexts, including other species and the physical environment. From this perspective, ecolinguistics is considered an essential trend in the education field, as it aims to raise student awareness towards dealing with nature and the environment.

The employment of ecolinguistics in the EFL classroom is theoretically guided by the theory of ecolinguistics that views language and environment as two interrelated areas. Ecolinguistics is also in line with the contextual teaching and learning approach (CTL) that takes into consideration the correlation between the lesson content and the real world. Learners need meaningful learning and critical thinking to be able to use language naturally in various situations. Contextual learning implies using concepts that are related to the context for the purpose of providing 'meanings, relevance, and benefites towards learning'

(Luardini & Sujiyani, 2018, p. 377; Fadel & Elyas, 2021). Within the scope of EFL learning, the contextual approach associates the content of the lesson and materials of language teaching to the real world to create meaningful learning (Nunan, 2003; 2015; Richards, 2006). Contextual learning makes language learning more natural, relevant and meaningful (Richards & Rodgers, 2014). Luardini and Sujiyani (2018) stated that 'When students know the relationships between the lesson in the classroom and real life, they realize the advantages of learning' (p. 376). Mühlhäusler and Peace (2006) extended the scope of ecolinguistics to study the 'cultural and political environment' (p. 8).

The integration of ecolinguistics in EFL contexts has several advantages that can be described from two perspectives, linguistics and ecology. From a linguistic perspective, motivation is the main advantage. The integration of ecolinguistics makes understanding concepts and the subject matter easier for students, as it is based on the natural, social and cultural surroundings (Waqar & Majeed, 2023; Faramarzi et al. 2021). This in turn motivates students to engage in learning activities and makes them feel more connected to the materials they use. Furthermore, to create integral meaning, 'learning language cannot be separated from the biological/natural, social, and ideological/cultural environments' (Tjendani et al., 2019, p. 257). Understanding the relationship between the classroom lesson and real life helps students realize the advantages of learning.

Some studies have been conducted to investigate the relationship between motivation and ecolinguistic integration. For example, Waqar and Majeed (2023) conducted a qualitative study to explore students' motivation when developing an ecolinguistic perspective-based syllabus.

From an ecological perspective, ecolinguistics is linked to environmental education, extending beyond teaching methods to enhance learners' environmental awareness and foster a sense of responsibility toward the environment. (Pooley & O'Connor, 2000). Ecological awareness in EFL/ESL contexts requires ecological pedagogy that relies on the employment of language in a way that contributes to ecological topics such as environmental sustainability, justice and well-being. Increasing learners' knowledge means changing learners' behaviour to consider their responsibility towards the environment by protecting the environment and natural resources. This can be done by implementing ecological practices in the classroom such as providing various ecolinguistics-based learning materials, activities, topics, and so on.

Although it may be a bit challenging for EFL/ESL teachers to integrate environmental awareness into the classroom due to political or institutional issues, several language experts such as Brown (2014) argued that one of the objectives in teaching is to help our students 'become informed about as many issues as possible that intrinsically affect their lives' (cited in Asgher et al., 2021, p. 706). Asgher et al., (2021), draws attention to the role of language teachers in increasing environmental education, stating that 'Positive attitudes and sustainable behavioural development is possible only with systematic and planned education and with teachers who have environmental awareness' (p. 707). Likewise, Halliday (1990) pointed out that maintaining a sustainable future is not only the responsibility of chemistry or biology teachers, but it is also the responsibility of language teachers, especially if the notion of sustainability is within the policy of an institute. Delavan (2020) also emphasized the role of language educators in raising students' ecological awareness and empowering them to shape a more sustainable future. However, having experience in this field is not potentially possible unless teachers are well-trained to acquire knowledge and experience about ecolinguistics.

Regarding the integration of ecolinguistics, the important question is how to integrate an ecolinguistic approach in EFL/ESL course design. The literature review shows that it could be integrated across all components of a course design such as content, materials, tasks and activities. Erten (2021) stated that environmental education should not be limited to the content of certain courses; rather, it could be interwoven into the content of all courses in the curriculum. For example, Asgher et al. (2021) reported that to enhance environmental literacy or education, it is essential to develop activities based on various learning domains, including cognitive (knowledge and skills), affective determinants of behaviour (attitudes, beliefs, emotions and the like) and psychomotor (physical activities).

Based on a UNESCO report (2016), ecolinguistics is considered part of education for sustainable development (ESD) that aims to expand learners' competencies and attitudes to deal with environmental issues (UNESCO, 2010). The United Nations has established six objectives for environmental education: (1) environmental issues are well-known; (2) a fundamental grasp of the environment, its issues and humans' role in the ecosystem; (3) a caring attitude towards environmental issues; (4) environmental problem-solving abilities; (5) the ability to assess potential environmental remedies; and (6) actual participation in carrying out solutions for environmental issues (UNESCO UNEP, 1976, cited in Jacobs & Goatly, 2000). One of the goals of environmental education is to generate ecologically literate citizens (Çakır et al., 2010).

Research methodology

The survey in this research was designed and shaped by a quantitative approach in collecting and analysing the data. The study employed a questionnaire that consists of two parts. The first part consists of seventeen statements about students' perspectives regarding the proposed idea of incorporating the ecolinguistics approach into the existing EFL existing curriculum. The second part consists of sixteen items investigating the methods and strategies of integrating ecolinguistics into the EFL curriculum. According to Fraenkel and Wallen (2006), survey research is 'a collection of information from a sample by asking questions to describe some aspects of the population of which the sample is a part' (p. 423). The data were collected by distributing questionnaires virtually through sharing a link to Google Forms with the respondents. The questionnaire was filled out by the respondents based on the provided instructions by the researchers. The study employed descriptive statistical techniques for the data analysis.

Sample

The sample of the work consists of 200 students at the foundation level at the University of Technology and Applied Sciences in Nizwa, Oman. The selection of the sample was based on their lived experiences and familiarity with the EFL curriculum, more specifically, their familiarity with the materials selection, activities and classroom practices. Choosing participants having experience in the subject matter, in this case, the topic that comprises this chapter, will potentially contribute to supporting the purpose of presenting comprehensive and meaningful findings (Denzin & Lincoln, 2008; Dornyei, 2012; Punch, 2013). In addition, the sample represented the targeted population in terms of the context where the research takes place and the objectives of the study (Introduction). Prior to distributing the questionnaire, the researchers delineated the purpose of the study and the items via emails and informal meetings.

Instruments of the data analysis

The questionnaire data were analysed and displayed descriptively using statistical features such as frequency counts and percentages. Google Forms software was used to analyse the data quantitatively online. The responses were automatically recorded and represented in a Google spreadsheet in an analysable format using graphical representations such as bar and pie charts. The analysis includes detailed responses in terms of percentages for each questionnaire item.

Findings

This section presents the findings obtained from the data analysis with the purpose of providing answers to the research questions. Based on the research questions and objectives, the findings are organized into two parts, Section One and Section Two findings. To illustrate the data quantitatively, tables including items and percentages are used along with the discussion.

Section One findings: Effects of ecolinguistics integration on students' motivation

This section is concerned with exploring the significance of integrating ecolinguistics into the existing curriculum in raising the students' motivation. It is related to the first research question, 'How does the proposed integration of ecolinguistics affect students' motivation and engagement in learning from the perspectives of students in the UTAS foundation programme?' In response to this question, the questionnaire survey included a set of items related to the pedagogical significance of the employment of ecolinguistics in the classroom, and the percentages of the respondents' answers comprise the first finding. Table 3.1 presents the findings according to each item with the percentages.

Overall, the statistics in Table 3.1 demonstrate that most of the respondents are in favour of the integration of ecolinguistics into the EFL curriculum. What follows is a detailed description of the findings according to each questionnaire item.

Ecolinguistics is important in English-language teaching

Table 3.1 shows that a large percentage of students strongly agreed (37 per cent) and agreed (30.3 per cent) that ecolinguistics is important in EFL teaching and learning, while only 1 per cent of the respondents disagreed, and 10.1 per cent felt that ecolinguistics was not important. Around a quarter of the participants (26.3 per cent) remained neutral.

Table 3.1 The Effects of Integrating Ecolinguistics on Students' Motivation

	Item	Strongly Agree (%)	Agree (%)	Neutral (%)	Disagree (%)	Strongly Disagree (%)
1	I believe that ecolinguistics is important in English-language teaching.	37	30.3	26.3	1	10.1
2	I believe that environmental education should be incorporated into language skills.	25.5	38.8	20.4	7.1	6.1
3	Ecolinguistics helps me to be a good citizen and have a positive impact on my country.	42.9	29.6	17.3	5.1	4.1
4	By using ecolinguistics approach, I learn English to represent the nature or environment I deal with every day.	36.5	25	24	7.3	5.2
5	It is important to know the relationship between the lesson in the classroom and real life outside the classroom.	25.3	39.4	23.2	7.1	2
6	I believe that ecolinguistics supports culture-based education.	35.7	28.6	21.4	8.2	3.1
7	Ecolinguistics increases my classroom engagement and school motivation.	34.7	25.5	24.5	7.1	6.1
8	Implementing ecolinguistics in EFL classroom helps me to raise my ecological awareness.	33.3	28.1	20.8	8.3	6.3
9	Ecolinguistics teaches me to love the environment and preserve the natural and cultural heritage of my country.	33	36.1	18.6	4.1	5.2
10	Ecolinguistics helps me develop my language skills on the one hand and become aware of the environment on the other hand.	53.6	11.3	25.8	4.1	3.1
11	Ecolinguistics improves my language skills.	36.1	23.7	26.8	8.2	3.1

(Continued)

Table 3.1 (Cobtinued)

	Item	Strongly Agree (%)	Agree (%)	Neutral (%)	Disagree (%)	Strongly Disagree (%)
12	I believe that it is important to learn societal environmental responsibility as part of education for sustainable development.	**32.7**	**35.7**	**15.3**	**8.2**	**6.1**
13	Introducing environmental linguistic expressions will build my vocabulary.	**56.8**	**2.1**	**17.9**	7.4	**5.3**
14	Ecolinguistics helps me develop higher mental abilities or skills.	**41.8**	**25.5**	**22.4**	**5.1**	**2**
15	Topics on environment and nature increase my motivation in the classroom.	**39.8**	**22.4**	**21.4**	**12.2**	**3.1**
16	Topics on environment and nature increase my engagement in classroom activities.	**35.1**	**25.8**	**24.7**	7.2	**5.2**
17	Topics on environment and nature increase my communication skills.	**36.1**	**24.7**	**23.7**	7.2	**4.1**

The integration of environmental education into language skills

With regard to the integration of environmental education into language skills, Table 3.1 illustrates that almost two-thirds of the respondents strongly agreed and agreed (25.5 per cent and 38.8 per cent, respectively), with the idea. In contrast, very few respondents disagreed and strongly disagreed (7.1 per cent and 6.1 per cent, respectively) with the idea, whereas the others (20.4 per cent) remained neutral.

Ecolinguistics helps students to be good citizens who have a positive impact on the country

As seen in Table 3.1, the majority of the respondents strongly agreed (42.9 per cent) and agreed (29.6 per cent) on the significance of ecolinguistics in motivating them to be good citizens with the expectation of having good impacts on the country. In comparison, only 5.1 per cent and 4.1 per cent reported their disagreement and strong disagreement, respectively. Those remaining neutral were 17.3 per cent.

Ecolinguistics helps students to learn English to represent the nature they deal with

As revealed in Table 3.1, around half of the respondents strongly agreed (36.5 per cent) and agreed (25 per cent). In comparison, a few respondents disagreed and strongly disagreed (7.3 per cent and 5.2 per cent, respectively). Around a quarter of the respondents (24 per cent) remained neutral.

The relationship between the classroom lesson and the outside classroom real life

The relationship between the classroom lesson and the outside classroom real life is elucidated in Table 3.1, which illustrates that over half of the respondents strongly agreed and agreed (25.3 per cent and 39.4 per cent, respectively). However, very few of them disagreed and strongly disagreed (7.1 per cent and 2 per cent, respectively).

Ecolinguistics supports culture-based education

According to Table 3.1, more than half of the respondents were between strongly agree (35.7 per cent) and agree (28.6 per cent) regarding the advantages of ecolinguistics in supporting culture-based education. In contrast, few respondents were between agree (8.2 per cent) and strongly disagree (3.1 per cent). Several students (21.4 per cent) were neutral.

Ecolinguistics increases classroom engagement and motivation

The relationship between ecolinguistics and classroom engagement and motivation is reflected in Table 3.1, which indicates that more than half of the respondents strongly agreed (34.7) and agreed (25.5), respectively. However, only 7.1 per cent and 6.1 per cent reported their disagreement and strong disagreement, respectively. Around 24.5 per cent remained neutral.

Implementing ecolinguistics in EFL classrooms raises ecological awareness

The implementation of ecolinguistics in EFL classrooms is explained through the statistics in Table 3.1, which shows the respondents' strong agreement (33.3 per cent) and agreement (28.1 per cent) in comparison to the percentages of disagreement (8.3 per cent) and strong disagreement (6.3 per cent). However, less than a quarter of the respondents (20.8 per cent) were neutral.

Ecolinguistics teaches students to love the environment and the heritage of the country

Similar to the previous item, Table 3.1 indicates that more than half of the respondents strongly agreed (33 per cent) and agreed (36.1 per cent) on the idea, whereas only 4.1 per cent and 5.2 per cent disagreed and strongly disagreed, respectively. The other respondents (18.6 per cent) remained neutral.

Ecolinguistics helps students develop their language skills and awareness at the same time

Table 3.1 shows that almost half of the respondents strongly agreed (53.6 per cent) and agreed (11.3 per cent). In contrast, few respondents disagreed (4.1 per cent) and strongly disagreed (3.1 per cent). About a quarter of the respondents (25.8 per cent) remained neutral.

Ecolinguistics improves my language skills

Table 3.1 indicates that more than half of the respondents (36.1 per cent and 23.7 per cent, respectively) strongly agreed and agreed on the role of implementing ecolinguistics in enhancing students' language skills. Less than one-third of the respondents (26.8 per cent) remained neutral, while the table shows that only 8.2 per cent and 3.1 per cent of the respondents disagreed and strongly disagreed, respectively, on the role of ecolinguistics in this regard.

The importance of education for sustainable development

The results for this item are similar to those of the previous one, as most of the respondents voted for the strong agreement (32.7 per cent) and agreement (35.7 per cent) options. Few respondents, on the other hand, voted for the disagreement (8.2 per cent) and strong disagreement (6.1 per cent) options. Several respondents (15.3 per cent) remained neutral.

Introducing environmental linguistic expressions will build students' vocabulary

Table 3.1 shows the respondents' strong agreement (56.8 per cent) and agreement (2.1 per cent) towards this item. However, few respondents showed their disagreement (7.4 per cent) and strong disagreement (5.3 per cent), while other respondents remained neutral (17.9 per cent).

Ecolinguistics helps students develop higher mental abilities

The results indicate that around two-thirds of the responses went for the options of strong agreement (41.8 per cent) and agreement (25.5 per cent) in comparison to the options of disagreement (5.1 per cent) and strong disagreement (2 per cent). The neutral options were around 22.4 per cent.

With regard to topics on environment and nature increasing motivation in the classroom, the results also indicate the respondents' strong agreement (39.8 per cent) and agreement (22.4 per cent) regarding the role of integrating ecolinguistics in raising students' motivation.

Likewise, the findings for Item 16, which asked whether environmental topics enhance student engagement, show that a majority of respondents either strongly agreed (35.1 per cent) or agreed (25.8 per cent). In contrast, 24.7 per cent disagreed and (7.2 per cent) strongly disagreed. A smaller portion (5.2 per cent) remained neutral.

Topics on environment and nature increase communication skills

The obtained results reveal the options of strong agreement (36.1 per cent) and agreement (24.7 per cent) in comparison to the options of disagreement (7.2 per cent) and strong disagreement (4.1). The neutral responses were 23.7 per cent.

Section Two findings: Methods and strategies for integrating ecolinguistics into the EFL curriculum

This section provides answers to the second research question: 'What strategies and methods must be implemented when integrating ecolinguistics into the EFL curriculum from the perspectives of students?' Table 3.2 presents the results in relation to every item.

Incorporating ecolinguistics through materials

This item is related to exploring the respondents' perspectives regarding the role of materials in implementing ecolinguistics into the EFL curriculum. Table 3.2 shows that 24.5 per cent of the respondents showed strong agreement, and 36.7 per cent of them showed agreement, whereas only 10.2 per cent and 7.1 per cent expressed their disagreement and strong disagreement, respectively. Around 21.4 per cent of them remained neutral.

Table 3.2 Methods and Strategies of Integrating Ecolinguistics into the EFL Curriculum

	Item	Strongly Agree (%)	Agree (%)	Neutral (%)	Disagree (%)	Strongly Disagree (%)
1	Ecolinguistics should be introduced into the EFL curriculum through materials such as textbooks, handouts, or PowerPoint presentations.	24.5	36.7	21.4	10.2	7.1
2	Ecolinguistics should be introduced into the EFL curriculum through activities.	28.1	34.4	21.9	9.4	6.3
3	Ecolinguistics should be introduced into the EFL curriculum through extra curriculum activities outside the classroom.	32	33	24.7	5.2	5.2
4	I prefer field trips to learn about language and environment.	32.7	40.8	15.3	9.2	2
5	I need workshops on language and environment.	36.7	33.7	16.3	9.2	4.1
6	I need training on language and environment.	36.5	32.3	21.9	5.2	4.2
7	I need visual aids on language and environment to learn more about ecolinguistics.	29.6	40.8	16.3	9.2	4.1
8	I need videos on environment to learn about ecolinguistics.	25.5	42.9	18.4	4.1	9.2
9	I believe the instructional design must be related to our native environment and culture.	29.6	35.7	23.5	8.2	3.1
10	I believe the instructional design must be related to global environment and culture.	39.8	30.6	21.4	3.1	5.1
11	I believe the instructional design must be related to both local and global environment and culture.	38.8	29.6	21.4	6.1	4.1

(Continued)

Table 3.2 (Continued)

	Item	Strongly Agree (%)	Agree (%)	Neutral (%)	Disagree (%)	Strongly Disagree (%)
12	I need more vocabulary on environment and nature, and global issues.	37.1	28.9	22.7	6.2	5.2
13	I need a taxonomy including definitions of concepts and themes related to environment and nature, and global issues.	32	35.1	21.6	5.2	6.2
14	I need narrative and descriptive texts on environment and nature, and global issues.	31.6	40.8	16.3	8.2	3.1
15	I need to be involved in projects on environment and nature.	31.6	36.7	20.4	6.1	4.1
16	I need to read about case studies related to environment and nature.	33.7	32.7	21.4	5.1	7.1

Incorporating ecolinguistics through activities

With regard to the incorporation of ecolinguistics through activities, the results in Table 3.2 also show similar results to Item 1, revealing that more than half of the respondents reported their strong agreement (28.1 per cent) and agreement (34.4 per cent). However, a few respondents reported their disagreement (9.4 per cent) and strong disagreement (6.3 per cent).

Incorporating ecolinguistics through curriculum activities

In the same vein, Table 3.2 reveals the respondents' strong agreement (32 per cent) and agreement (33 per cent) towards the importance of extracurricular activities in implementing ecolinguistics. In comparison, a few respondents (5.2 per cent and 5.2 per cent, respectively) reported their disagreement and strong disagreement. Around (24.7 per cent) of the respondents remained neutral.

Incorporating ecolinguistics through field trips

According to Table 3.2, over two-thirds of the respondents (32.7 per cent and 40.8 per cent) reported strong agreement and agreement regarding their preferences for field trips in teaching ecolinguistics. However, only 9.2 per cent and 2 per cent reported disagreement and strong disagreement, respectively. Several students (15.3 per cent) remained neutral.

The need for workshops on language and environment

The results for this item indicate that more than half of the respondents (36.7 per cent and 33.7 per cent, respectively) strongly agreed and agreed on the need for the workshops in this regard. On the other hand, a few respondents (9.2 per cent and 4.1 per cent, respectively) reported their disagreement and strong disagreement. Those remaining neutral were 16.3 per cent.

Training on language and environment

Table 3.2 also shows similar results to the previous items, as most of the responses were between strongly agree (36.5 per cent) and agree (32.3 per cent), while few responses were between disagree (5.2 per cent) and strongly disagree (4.2 per cent). Less than a quarter of the responses were neutral.

Incorporating ecolinguistics through visual aids

The statistics in Table 3.2 reveal that most of the responses went for the options of strongly agree (29.6 per cent) and agree (40.8 per cent) compared to the responses that went for disagree (9.2 per cent) and strongly disagree (4.1 per cent). The responses that were neutral were 16.3 per cent.

Incorporating ecolinguistics through videos

Similar to the previous item, most of the respondents reported their strong agreement (25.5 per cent) and agreement (42.9 per cent), while few respondents reported their disagreement (4.1 per cent) and strong disagreement (9.2 per cent).

Relating ecolinguistics to students' native environment and culture

Table 3.2 also shows similar results to the previous items, revealing the respondents' strong agreement (29.6 per cent) and agreement (35.7 per cent) on the item. Very few respondents, on the other hand, expressed their disagreement (8.2 per cent) and strong disagreement (3.1 per cent).

Relating ecolinguistics to the global environment and culture

According to Table 3.2, more than two-thirds of the responses went for the options of strong agreement (39.8 per cent) and agreement (30.6 per cent) agreement in contrast to the few responses that went for disagreement (3.1 per cent) and strong disagreement (5.1 per cent). Those who remained neutral were about (21.4 per cent). According to the analysis of Item 11, which asked whether students should read case studies related to the environment, the results presented in Table 3.2 indicate that the majority of students either strongly agreed (33.7%)

or agreed (32.7%) with the statement. In contrast, a small percentage expressed disagreement (5.1%) or strong disagreement (7.1%). Approximately a quarter of the students (21.4%) remained neutral.

The need for vocabulary on environment and global issues

The results obtained reveal that many respondents strongly agreed (37.1 per cent) and agreed (28.9 per cent) with the statement, while only a few respondents disagreed and strongly disagreed (6.2 per cent and 5.2 per cent, respectively). Less than a quarter of the responses were neutral (22.7 per cent). Regarding the need for a taxonomy including definitions and concepts related to the environment, the obtained results also show the respondents' strong agreement (32 per cent) and agreement (35.1 per cent), while few responses went for the options of disagreement (5.2 per cent) and strong disagreement (6.2 per cent). The other respondents (21.6 per cent) remained neutral.

The need for narrative and descriptive texts on global issues

Table 3.2 demonstrates that more than two-thirds of the respondents voted for the strongly agree (31.6 per cent) and agree (40.8 per cent) options, in contrast to the options of disagreement and strong disagreement, which received lower percentages (8.2 per cent and 3.1 per cent, respectively).

With regard to the need to be involved in projects on the environment, the obtained results indicate that respondents strongly agreed (31.6 per cent) and agreed (36.7 per cent). Few respondents voted for the disagree and strongly disagree options (6.1 per cent and 4.1 per cent, respectively). The other respondents (20.4 per cent) remained neutral. Regarding the last item about the need to read case studies on the environment, the results reveal that more than half of the respondents were between strongly agree (33.7 per cent) and agree (32.7 per cent), while few of them were between disagree (5.1 per cent) and strongly disagree (7.1 per cent). Those remaining neutral were 21.4 per cent.

Discussion of the results

The purpose of this chapter was to explore the perceptions of EFL students at the foundation programme regarding whether the integration of ecolinguistics into the existing curriculum would affect their motivation and engagement in the classroom. It also explored their perceptions regarding the methods and strategies that should be utilized when integrating ecolinguistics in the classroom. As shown in the previous section, the findings suggest that the majority of respondents agreed with the proposed idea, and provide a set of suggestions regarding the methods and ways of integration.

The findings in Section One provided insight into the first research question: 'How does the proposed integration of ecolinguistics affect students' motivation and engagement in learning from the perspectives of students in the UTAS foundation programme? The findings reveal that most of the respondents appreciated the proposed idea of integrating concepts, topics and themes related to ecolinguistics into the existing curriculum. This finding implies their awareness of environmental education as a new trend in education that aligns with global issues such as global warming, flooding, air/water pollution and the like. Acknowledging the importance of ecolinguistics or environmental education is considered a strong motivation that may assist and accelerate the learning process. Accordingly, they assigned higher percentages to the need to integrate ecolinguistics into the EFL classroom.

The findings in Section One coincide with the literature review and previous studies (such as Tjendani, 2019; Waqar & Majeed, 2023) that recognized the importance of integrating ecolinguistics into the EFL context for the purpose of raising students' motivation. The findings also coincide with those of several studies that have been compiled by the US National Environmental Education Foundation (NEEF, www.neefusa.org), which indicated that implementing environmental education fosters students' critical thinking skills, motivation and classroom interaction.

Motivation is a crucial factor in EFL classrooms (Al-Khalidi, 2020, 2019; Dörnyei, 1994). Raising students' motivation is determined by certain pedagogical practices, and among them is considering the relationship between language and the real world (Richards, 2006; Brown, 2014). Environmental education is part of contextual teaching and learning that targets meaning as a key element in the learning process (Richards, 2006; Johnson, 2017).

Section Two represented the responses to the second research question: 'What strategies and methods must be implemented when integrating ecolinguistics into the EFL curriculum from the perspectives of students?' The obtained findings emphasized the role of materials, syllabus or content, and activities when integrating ecological texts, concepts and themes into EFL lessons. Students believed that the integration of ecolinguistics through these components would potentially create classroom interaction, motivation and engagement. Based on the obtained results, the researchers can draw several conclusions. First, teachers who are more concerned about students' motivation and involvement need to be aware of the selection of materials and content. As a matter of fact, not all teachers are aware of ecolinguistics and environmental education, as some believe that it is the responsibility of science teachers. However, this chapter and previous studies on ecolinguistics argue that EFL teachers need to be aware of this approach as a new trend within the scope of ELT. Second, teachers need to be aware of selecting and adapting materials based on learners' needs in accordance with the requirements of modern and global education (Al-Khalidi, 2023a, 2023b; Long, 2005). According to Asgher et al. (2021, p. 708), 'Focusing on the teacher education process is one of the most essential strategies to help teachers acquire environmental awareness and responsibility.' Also, according to UNESCO (2024), it has become crucial for institutions to consider environmental education.

The results of the chapter coincide with the previous findings by Waqar and Majeed (2023), who conducted a qualitative study to 'develop the ecolinguistic perspective-based syllabus to improve language learners' motivation leading to better language learning' (p. 103). Likewise, Wang (2019) conducted a study to apply ecolinguistic theory when teaching EFL in China. The study concluded that to regulate the learning process, students should be encouraged to select the content of the instructional materials. The flowchart in Figure 3.1 provides a framework outlining the findings obtained from the theoretical and practical views.

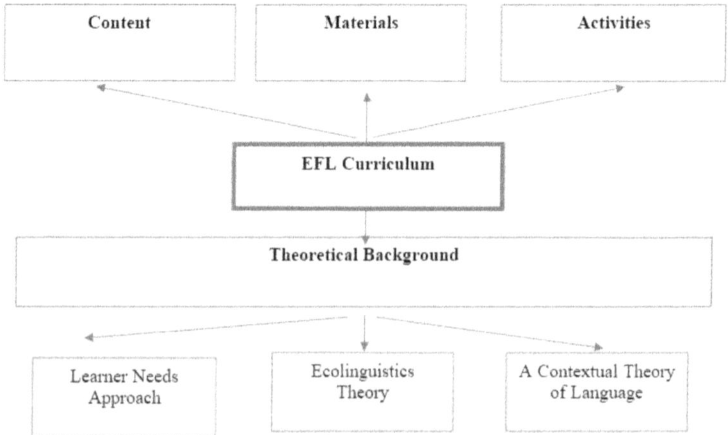

Figure 3.1 A framework of integrating ecolinguistics into the EFL curriculum.

Conclusion

This exploratory chapter investigated student perceptions of the anticipated effects of the integration of ecolinguistics into the EFL curriculum on students' motivation. The chapter was theoretically guided by the approaches of ecolinguistics, learner needs and contextual teaching that emphasize acknowledging the role of ecolinguistics to comply with the demands of modern education and the objectives of UNESCO. As far as the students' perceptions of integrating ecolinguistics into the existing curriculum with reference to motivation are concerned, the findings reveal that the proposed idea has the potential to optimize the learning process through motivation, understanding and ecological awareness. Therefore, it is recommended to integrate ecolinguistics into the entire curriculum through the effective selection of materials, activities, and syllabus or content.

This chapter also recommends a framework (as in the flowchart in Figure 3.1) that can be used to integrate ecolinguistics into EFL classes through course or lesson design. However, training for teachers, course designers and students should be carried out to consider new trends in EFL. Also, it is recommended to coordinate with ecologists and associations concerned with ecolinguistics such as The International Linguistics Association National Geographic when designing language courses to integrate ecolinguistics professionally and communicatively. National Geographic Learning, for instance, is a very useful organization that 'brings the world to the classroom and the classroom to life' (National Geographic Learning). Through its partnership with National Geographic, National Geographic Learning develops programmes in language and skills that learners need to be successful citizens and leaders. 'Pathways', for example, is a global, five-level academic English programme with materials from authentic content and guided lessons that aim to develop language skills, critical thinking and ecolinguistic awareness.

References

Al-Khalidi, I. (2019). Investigating motivational factors in EFL classroom from the perspectives of students at a tertiary context in Oman. In S. Hidri (Ed.), *English language teaching research in the Middle East and North Africa: Multiple perspectives* (pp. 479–505). Palgrave Macmillan.

Al-Khalidi, I. (2020). Investigating the process of EAP course design in a college context from the perspectives of teachers. *British Journal of English Linguistics*, *8*(4), 1–26.

Al-Khalidi, I. (2023a). Investigating the process of developing a workplace English digital course for Oman petroleum academy and industry. *World Journal of English Language, 13*(7), 1–17.

Al-Khalidi, I. (2023b). Investigating the integration of a workplace English digital course into the curriculum of the university of technology and applied sciences in Oman. *Journal of Namibian Studies, 33*(3), 345–374.

Asgher, T., Saleem, M., & Ilyas, A. (2021). Ecolinguistics: Integrating environmental education in teaching English as second language. *Pakistan Languages and Humanities Review, 5*(2), 705–712.

Brown, H. D. (2014). Principles of language learning and teaching: A course in second language acquisition. Pearson.

Çakır, M., Serhat İ., & Özgür, K. Doğan. (2010). Understandings of current environmental issues: Turkish case study in six teacher education colleges." *Educational Studies* 36(1): 21–33.

Delavan, M. G. (2020). Earth democracy as empowerment for TESOL students and educators: Though the crisis speaks English, Englishers can become a commons language of sustainability. In Jason Goulah & John Katunich (Eds.), *TESOL and sustainability: English language teaching in the Anthropocene era* (pp. 19–40). Bloomsbury Publishing.

Denzin, N. K., & Lincoln, Y. S. (2008). Introduction: The discipline and practice of qualitative research. In N. K. Denzin & Y. S. Lincoln (Eds.), *Strategies of qualitative inquiry* (3rd ed., pp. 1–43). Sage Publications.

Dörnyei, Z. (1994). Understanding L2 motivation: On with the challenge. *The Modern Language Journal, 78*(4), 515–523.

Dörnyei, Z. (2012). *Research methods in applied linguistics*. Oxford University Press.

Environmental Education Foundation (NEEF). www.necfuss.org

Erten, S. (2021). *Different perspectives on environmental education*. ISER Publishing.

Fadel, S., & Elyas, T. (2015). ESP needs analysis to integrate a scientific reading program in the English language institute at King Abdul-Aziz University. *International Journal of Educational Investigations, 2*(4), 14–27.

Faramarzi, A., Janfeshan, K., & Ars, A. (2021). An investigation into ecolinguistics representation in Iranian high school English textbooks. *Ars Artium: An International Refereed Research Journal of English Studies and Culture, 9*(10), 70–85.

Fill, A. F., & Penz, H. (2018). Ecolinguistics in the 21st Century: New orientations and perspective. In A. F. Fill & H. Penz (Eds.), *The Routledge handbook of ecolinguistics* (pp. 437–443). Routledge.

Fraenkel, J., Wallen, N., & Hyun, H. (1993). *How to design and evaluate research in education* (10th ed.) McGraw-Hill Education.

Fraenkel, J., Wallen, N., & Hyun, H. (2011). How to design and evaluate research in education (10th ed.) McGraw-Hill Education.

Halliday, M. (1990). New ways of meaning: A challenge to applied linguistics. *Applied Linguistics, 6*, 7–36.

Jacobs, G. M., & Goatly, A. (2000). The treatment of ecological issues in ELT coursebooks. *ELT Journal*, 54, 256–264. 77.

Jacobs, G. M., Jiexin, T., & Michael, J. (2016). The presentation of animals in English as an additional language coursebooks. *Online Submission.* https://files.eric.ed.gov/fulltext/ED574076.pdf

Jarvis, H., & Atsilarat, S. (2005). Shifting paradigms: From a communicative to a context-based approach. *English Language Teaching Journal, 50*(1), 9–15.

Johnson, K. (2017). An introduction to foreign language learning and teaching. Routledge.ntext-based approach. *English Language Teaching Journal*, 50(1), 9–15.

Long, M. H. (2005). Methodological issues in learner needs analysis. In M. H. Long (Ed.), *Second language needs analysis* (pp. 19–75). Cambridge University Press. https://doi.org/10.1017/CBO9780511667299.002

Luardini, M., & Sujiyani, E. (2018). Introducing ecolinguistics in EFL classroom. In *international conference on teacher training and education* (pp. 375–382). Atlantis Press.

Mühlhäusler, P., & Peace, A. (2006). Environmental discourses. *Annual Review of Anthropology, 35*, 457–479.

Nunan, D. (2003). Methodology. In David Nunan (Ed.), *Practical English language teaching*. McGraw Hill.

Nunan, D. (2015). *Teaching English to speakers of other languages: An introduction.* Routledge.

Pooley, J. A., & O'Connor, M. (2000). Environmental education and attitudes: Emotions and beliefs are what is needed. *Environment and Behavior, 32*(5), 711–723.

Punch, K. F. (2013). *Introduction to social research: Quantitative and qualitative approaches*. Sage.

Richards, J. C. (2006). *Communicative language teaching today.* Cambridge University Press.

Richards, J. C., & Rodgers, T. S. (2014). *Approaches and methods in language teaching.* Cambridge University Press.

Stibbe, A. (2021). *Ecolinguistics: Language, ecology and the stories we live by* (2nd ed.). Routledge. ISBN: 978-0-367-42841-9

Tjendani, E. N., Suastra, I. M., Mbete, A. M., & Seri Malini, N. L. N. (2019). An ecolinguistics perspectives for English syllabus development. E. *Journal of Linguistics, 13*(2), 249.

UNESCO. (2024). The UNESCO recommendation on education for peace, human rights and sustainable development: An explainer. Retrieved from: https://unesdoc.unesco.org/ark:/48223/pf0000388330

UNESCO. (2010). Education for sustainable development: An expert review of processes and learning. Paris: UNESCO. https://unesdoc.unesco.org/ark:/48223/pf0000186829

UNESCO. (2016). Education for people and planet: Creating sustainable futures for all. Global Education Monitoring Report 2016. Paris: UNESCO. https://unesdoc.unesco.org/ark:/48223/pf0000245752

Waqar, T., & Majeed, M. (2023). An eco linguistic approach to EFL teaching: A case study of multilingual language learners in Pakistan. *Journal of Positive School Psychology*, *7*(1), 103–153.

Wang, L. (2019, January). Research on English classroom teaching from the perspective of ecological Linguistics. In 3rd International Seminar on Education Innovation and Economic Management (SEIEM 2018) (pp. 153–156). Atlantis Press.

Zahoor, M., & Janjua, F. (2020). Green contents in English language textbooks in Pakistan: An ecolinguistic and Eco pedagogical appraisal. *British Educational Research Journal*, *46*(2), 321–338.

Zhou, W. (2022). Ecolinguistics: A half-century overview. *Journal of World Languages*, *7*(3), 461–486.

4

Children's books for enjoyment, language learning and the Sustainable Development Goals

Chenghao Zhu, Meng Huat Chau, Hisham M. El-Tahawy, Jasper Roe and George M. Jacobs

Introduction

Adults have been telling stories to children since the early days of language, not to mention children telling stories to adults, to other children and to themselves. With the invention of printing many centuries ago, adults' stories for children could be made into books (Leland et al., 2017). With further advances in technology, stories for children can now be shared via other media, as well as multimedia. Literature, both fiction and non-fiction, is available for children of all ages, even babies and toddlers. At the other end of the birth-to-early-adult age spectrum, books for teens and young adults attract many readers, listeners and viewers. Some adults, including academics, view children's literature as inferior to literature for adults. However, Le Guin (1992), a widely acclaimed author who wrote for both adults and youth, countered by sarcastically stating, 'Sure it is simple, writing for kids / Just as simple as bringing them up' (p. 49).

Children's literature, especially for preteens, has been recommended as a source of reading materials for second-language learners (Ahlquist, 2023). Cho and Krashen (2016) reported that the Sweet Valley series of children's books appealed to and benefitted people acquiring English as a second language. However, it should be noted that Webb and Macalister (2013) found children's books less useful than graded readers, that is, books created especially for second-language learners, due to the greater vocabulary difficulty of the children's books.

Researchers study children's literature (e.g. Hintz & Tribunella, 2019; Ismail, 2023), and university courses examine it. Ecolinguistics, which 'explores the role of language in the life-sustaining interactions of humans, other species and the physical environment' (International Ecolinguistics Association, 2023), presents researchers and the general public, including students and teachers, with tools and perspectives that can be used to better understand children's literature and other texts (i.e. collections of language) of any kind: spoken, written, visual or multimedia. The specific approach to ecolinguistics employed here derives from the work of Stibbe (2021). The website of the International Ecolinguistics Association offers a free online asynchronous course based on Stibbe's ideas.

Ecolinguists often label the view promoted by the powerful in society as anthropocentrism (*anthro* means human), that is, human-centred perspectives, caring only about the needs of humans and, more narrowly, the wealthier segment of the human population, those who use far more than their fair share of the Earth's resources and suffer the least from the consequences of the climate crisis and other ills. In contrast to anthropocentrism, ecolinguists promote ecocentrism, that is, perspectives representing the interests of all the planet's inhabitants with special attention to non-humans and less individually powerful humans. Thus, ecolinguistics seeks not only to understand but also to use that understanding to promote and guide actions for a greener, more just world.

Ecolinguistics also asks how texts, including children's literature, might affect and be affected by humans' beliefs and actions towards the environment. Penz and Fill (2022) explored the roots and branches of ecolinguistics. The branch utilized in the current article is based on the work begun by Halliday (1990), who called linguists' attention to the role played by language in the negative impacts that humans are having on the environment and all the planet's inhabitants. At the same time, ecolinguistics encourages people to identify and popularize texts that highlight pro-environment thoughts and actions. This is a key goal of this chapter.

The next section of the chapter describes a set of tools developed by ecolinguists and others: nine categories of stories told to readers. These story categories are not typical stories; they are mental frameworks for viewing the world. When readers understand these stories, they can become more critical readers. Next, the article recounts the background and plot of a popular children's book set in India. Then, some of the nine stories and other tools are used to critically explore the book in the hope that students and teachers will be able to apply these same tools to think critically about other texts, regardless of whether they are books, films or other media.

Literature review

Nine types of stories

Stibbe (2021) talked about the stories in our minds that impact our thoughts and actions. The term *story* here represents a valuable but somewhat unique use of the term. Stibbe detailed nine types of stories as described below.

1. Ideology – This refers to the beliefs individuals and groups of people have; for example, some people tend to believe that their personal value is determined by the amount and cost of their possessions (Zhang et al., 2020). This belief promotes anthropocentric ideology (the view that humans are the most important of the Earth's inhabitants; therefore, humans are right to ignore the effect of human actions on the Earth's other inhabitants), whereas others tend to believe that people's value is determined by what they do (such as helping others) and what their skills are (such as knowing how to repair machinery). This belief promotes ecocentric ideology (the view that all the Earth's inhabitants each have their unique value).
2. Framing – This refers to looking at a situation through the lens, also known as a frame, of another situation. For example, is the climate crisis a horrible problem that threatens the existence of *Homo sapiens* and other species, or is the climate crisis a gigantic opportunity for *Homo sapiens* to change from anthropocentric views and practices and, instead, adopt ecocentric views and practices? Thus, the question becomes whether to view the climate crisis through a problem frame or an opportunity frame.
3. Metaphors – Stories that describe something as if it were something else, that is, using a figure of speech to explain something to which it is not literally connected. For example, many people use the term 'Mother Earth', although everyone has/had a human mother. The Mother Earth metaphor can be ecocentric, encouraging humans to protect the environment as a form of repayment for the Earth providing for us, just as mothers do, and just as humans may feel an obligation to look after the women (not to mention men) who raised them. In contrast, an anthropocentric metaphor might be to use 'harvest', a term which normally collocates (i.e. appears in texts with) with crops, such as soybeans, to refer to the act of slaughtering non-human animals for meat. The harvest metaphor is anthropocentric because it denies the sentience (i.e. ability to think

and feel) of the animals whom we use for meat, eggs and dairy, thereby making their exploitation less objectionable.

4. Evaluations – Many stories evaluate the value or quality of ideas, species, groups of people, activities and so on. For example, do we protect people's right to use their money for almost anything they want, such as spending millions of dollars on a trip to outer space or hundreds of millions of dollars on a massive yacht (Koenig, 2023)? That would be an anthropocentric evaluation. Or do we use wealth produced via the planet's resources and many people's efforts and intelligence to provide sanitation systems for the billions who now live without such systems? In our evaluation, which right has priority: people's right to use whatever money they have accumulated, however they accumulated it, to do whatever they wish, or people's right to have the basics of a decent life, such as functioning sanitation systems?
5. Identities – This refers to stories that describe the characteristics of certain types of people. These characteristics could be the people's diets, clothing, appearance, actions and the language they speak and the way they speak it (i.e. language varieties, including regional, ethnic and social class varieties). For instance, what is the identity of successful people? Do stories of successful people highlight their individual accomplishments, or do they highlight these people's actions for the betterment of the community of which they are a part, including non-humans? For instance, the United Nations' Sustainable Development Goals (SDGs) (Lim et al., 2023; United Nations, 2023) identify characteristics of good people as including concern for those with less (e.g. SDGs 1 and 2) and for nature, both on land (e.g. SDG 15) and below water (e.g. SDG 14), as well as people who recognize and act on connections among the seventeen SDGs.
6. Convictions – These are stories about whether particular ideas, information or people are trustworthy. For example, an anthropocentric belief is that humans, even humans in the wealthy world, need large amounts of protein, and that the only reliable source of protein lies in meat, dairy and eggs from animals raised exclusively to provide food for humans. A healthier and ecocentric approach would be to trust the science that suggests that many people in the wealthy world have excessive protein intake, and that many alternative and sufficient sources of protein exist, for example, legumes, grains, nuts, seeds and plant-based meats, that do not harm non-human animals (Harvard Medical School, 2020).

7. Erasure – Stories of erasure are seen when beings or ideas are viewed as not worthy of mention or consideration. Many anthropocentric examples of erasure can be seen – for instance, photos of idyllic resorts with no mention of the environmental damage that took place in constructing the resorts or the living conditions of the people who live near the resorts. In contrast, ecocentric texts highlight these important features of reality.
8. Salience – Related to erasure, stories about the relative importance of beings or ideas involve the concept of salience. For example, Chau et al. (2022) compared the amount of newspaper coverage given to the COVID-19 pandemic with the amount of newspaper coverage given to the ongoing disasters of lack of food, clean water and sanitation. Even though, compared to COVID-19, more people die and are disabled due to the very solvable global shortages of food, clean water and sanitation, the four newspapers studied gave much more coverage to the pandemic, perhaps because the owners, producers and users of the newspapers felt that those dying from COVID were more important, more interesting and more like themselves, that is, more salient.
9. Narratives – Narratives are structures for organizing stories in a logical sequence of related settings and occurrences. An example of a narrative can be seen in the fictional animated children's film *Sea Beast* (Williams, 2022). The film presents an overall ecocentric perspective, as the film's two main protagonists begin the story attempting to hunt large fictional sea creatures (i.e. the so-called sea beasts of the film's title) to extinction. However, based on what they learn during the film, these two characters and many others switch to seeking successfully to protect the sea beasts and create harmony between the creatures and humans. Additionally, the sea beasts change their views of humans. In contrast, a famous anthropocentric example of a narrative about large marine beings is the novel *Jaws* (Benchley, 1974), later made into a famous Hollywood film. The novel and film tell the story of a shark who terrorizes humans and whom humans eventually kill.

An ecocentric children's novel

Stibbe's 2021 book is titled *Ecolinguistics: Language, Ecology and the Stories We Live By*, and too often, those stories promote anthropocentrism. In response, Ma and Stibbe (2021) issued a call to identify, produce and popularize stories that promote ecocentrism. Ma and Stibbe (2021) explained that ecocentric stories 'describe the world as web of sustaining relationships of humans with each other, other organisms, and the physical environment, and in doing so resist violent, prejudiced, and self-centred ways of seeing our world' (p. 186). Furthermore, Ma and Stibbe (2021, p. 186) stated that stories

> that work towards wellbeing and social justice for humans and care for the ecosystems that support life are the kind of stories which we need to promote and celebrate, so that our societies can be set on a fair, ethical, sustainable path out of social turmoil and ecological crisis.

Along similar lines, Martin (2004) proposed positive discourse analysis (PDA), which seeks to bring attention to 'heartening accounts' of 'social processes that make the world a better place' (p. 184). The current chapter examines a book (Bajaj, 2022) that may meet many of the criteria Ma and Stibbe proposed. *Thirst* is a popular book that was listed as a *New York Times* best-seller. The book is an English-language novel written for children about twelve years old. Before examining the book, this chapter summarizes the book's setting, characters and plot.

The setting, characters and the plot of *Thirst*

The novel *Thirst* highlights one of the SDGs (SDG 6), which involves the overlapping needs of clean water and sanitation. Every year, millions die due to the lack of safe drinking water and sanitation. Although progress has occurred, 2.2 billion people worldwide live without secure access to safe drinking water, and 4.2 billion people (more than half the world) lack safely managed sanitation services (UNICEF, 2019).

Thirst is a fictional story set in the slums of Mumbai, India, a city of approximately seven million. The characters in the story are not real, but their situation of water shortages for the poor is all too real, and, as the story begins, the climate crisis is making the situation worse, as it has shortened the monsoon season upon which the poor rely for some of their water. According to the book's author, while slum-dwellers constitute 40 per cent of Mumbai's population, the government allocates them only 5 per cent of the available water.

The story's protagonist is Minni, a twelve-year-old, living with her fifteen-year-old brother, Sanjay, and their parents in a Mumbai slum. Her mother takes care of the house and earns money as a maid, and her father has a tea shop in the neighbourhood. Theirs is a very tight-knit family, and the community in which they live is like what in Singapore and Malaysia is referred to as a kampung, a village in which everyone looks out for, roots for and supports everyone else. Minni excels at school. Her family must pay for the school she attends, which is reputed to be better than the public school where she would otherwise study.

Of course, Minni and the other people in the slum are poor, leading to multiple problems in such areas as jobs, education, health, sanitation and government resources. One of the biggest problems is water. Near Minni's home is a communal tap, but water is available only three hours a day, there are long queues and the water must be boiled for safety. Even worse is that sometimes little or no water comes from the tap. Meanwhile, the wealthy family for whom Minni's mom works as a maid lives not far away in a high-rise residence. There, water flows effortlessly from the taps found all over their apartment, including in the attached bathroom of the family's twelve-year-old daughter.

The slum's water shortage provides an opportunity for the 'Water Mafia', a real phenomenon in Mumbai. They steal water and sell it to people at high prices. Through a chance occurrence, Minni and her brother, along with each of their best friends, witness the Water Mafia stealing water, and one of the mafia leaders sees the two boys. They only escape because one of their slum neighbours is working for the mafia and, just in time, warns the boys to run away. As a result of Sanjay and his friend being seen by the Water Mafia leader, the families of both boys decide they must leave Mumbai and live indefinitely on the farm of one of the friend's family members to avoid being killed by the mafia. Indeed, the helpful neighbour is later murdered.

Soon after Minni's brother leaves Mumbai, her mother must leave too, as she has been having health problems possibly related to drinking unclean water. The family worries that she might have cancer. The mother decides to go to the countryside to stay with her mother and sister. There, the air is cleaner, and the mother will have less work to do. The mother's absence is clearly a plot device to facilitate readers seeing a fuller picture of the realities of slum life through Minni's eyes; for example, Minni takes over her mother's maid job. In general, in children's literature, the book's target audience is approximately the age of the book's protagonist. Thus, learning about life as a maid through Minni's eyes may have more impact on preteens than experiencing a similar phenomenon through the eyes of Rohini, Minni's mother.

Indeed, her mother's absence means that all of the work Rohini used to do now falls to Minni. This includes waking up early to join the queue to collect water and then boiling the water, not to mention making breakfast for herself and doing other family chores. Additionally, after school, Minni must temporarily take over her mother's job as a maid because that job is an important source of income for the family. Before her mother goes to the countryside, she explains to Minni how to do the work, part of which involves playing the submissive role of a servant, such as never talking unless spoken to by the mother and the mother-in-law of the family.

With all these new responsibilities, Minni starts arriving late at school and failing in her studies. Various people – including the school security guard, Minni's teacher, her best friend, Faiza, and Faiza's mother – help. For instance, when Minni misses school, Faiza takes detailed notes, and Faiza's mother makes food for Minni.

The employer family's twelve-year-old daughter, Pinky, is very friendly to Minni, and the two kids have fun together until Pinky's grandmother tells her to stay away from Minni because people from the slum are low caste and likely to carry diseases such as tuberculosis. Until near the end of her time as a maid, Minni never sees Pinky's father. When he does appear, he is very rude to Minni, who recognizes him as the Water Mafia leader she and Sanjay saw.

Minni wants the mafioso to be arrested so that her beloved brother Sanjay can safely return to Mumbai. Fortunately, although stories of police corruption are well-known in the slum, a new police leader has just taken over in their area. Minni and Faiza manage to send a note to him, exposing Pinky's father as a leader of the water thieves and informing the police where he lives. Pinky's father is arrested.

The doctor at the clinic where Minni's mom went before leaving for the countryside comes with good news that the mom has nothing more serious than hepatitis A, which is treatable. Her employer's family had all been vaccinated against the disease, but the vaccine is out of reach of the poor, although the doctor says that may change. The mom returns to Mumbai, but Sanjay decides to stay at the farm where he can live his dream of being a chef at a restaurant. Except for her brother's absence, all is almost back to normal for Minni, who is now taking a computer course with aspirations to escape the slum via a job in the computer industry.

Discussion

Thinking about *Thirst* via stories we live by

As explained above, Stibbe (2021) proposed nine overlapping categories of stories that could be used to think critically about texts, and three of these categories are used in the following analysis: erasure, ideology and metaphor. Critical thinking can be defined as analysing and evaluating. The evaluation can be positive, negative or some mix of the two. One type of story is erasure. Normally, people who live in slums are erased; that is, they do not appear in texts, as they are viewed as unimportant and uninteresting. However, in *Thirst*, slum-dwellers are not erased; they are featured, and readers see the world through their eyes.

Along similar lines, as to the story category of 'evaluations', Minni, her family and their neighbours are presented as good people, people who care about others, who work hard and who are intelligent. Although Pinky's grandmother disparages people in the slums, the book evaluates them positively. In contrast, Pinky's grandmother is depicted as a cruel person, who cares much more about status and material possessions than about human relationships. Not only is her thinking anthropocentric, but her prioritizing of humans is inclusive of only a narrow group of humans. Pinky's mother is shown as a basically nice person who is caught up in a story that leads her to spend much of her time shopping and some of the rest of it doing nothing more than following Minni around the home to make sure she does not miss any spots while cleaning. Pinky herself seems to be lost in life, unable to overcome the way her family isolates her from others, even from others of the same social class. In comparison with Minni, Pinky is inactive and not particularly smart.

As to 'ideology', another of Stibbe's overlapping stories, the lifestyle of those like Minni has a very small carbon footprint. They are the people who suffer from the ongoing climate crisis; for example, the water they need to live is becoming increasingly scarce. Thus, in some ways, Minni is both a climate crisis hero for her small carbon footprint and a victim of the crisis. However, is Minni really a hero given that her relatively minuscule footprint is involuntary? As will be explained later, she dreams of a life with a larger footprint. Of course, what characters in fictional stories do and think about are the inventions of the stories' authors.

The book does not seem to have a clearly ecocentric ideology, as the plight of other beings and the overall ecosystem receives little attention. However, justice may be the key point of the book's ideology. As noted above, Ma and Stibbe (2021, p. 186) called for 'a fair, ethical, sustainable path out of social turmoil and ecological crisis'. With the worsening of the climate crisis, increased attention has been placed on climate justice (Newell et al., 2021), that is, that the ill effects of income disparities in and between countries are exacerbated by the climate crisis, with the brunt of the impact being felt by those who contribute the least to global warming and other environmental harms. The book's central purpose seems to be to highlight the injustice of water shortages badly affecting the poor, while wealthier people treat water as a given. In this way, *Thirst* aligns well with SDG 6, 'Clean Water and Sanitation'.

Another issue with ideology concerns whether or not children's literature needs happy endings. Lim (2009) wrote a fictional children's book about the relationship between a preschool child and her grandmother. Near the end of Lim's story, the grandmother falls ill, goes to the hospital and dies. Tears are shed by the child and her parents. When asked why she gave the book a sad ending, Lim (personal communication) stated her belief that children, even very young children, should be exposed to the bumps in life.

Metaphor is another way that stories are told. The word *dream* in its various forms (e.g. *dreaming*) can be a metaphor for hopes and goals that people have. Twenty-three instances of *dream* appear in *Thirst*. All refer to the hopes of Minni, her brother Sanjay or their mother's hopes for her two children. For instance, early in the book, Minni says to Sanjay, 'One day, we'll live in one of those tall shiny buildings, where water runs from taps', to which Sanjay responds, 'Like the boy who was born here [the Mumbai slums] and studied computers and now has an office in a building and employs sixty people' (Bajaj, 2022, p. 3).

Dreams, circle of compassion and meritocracy

These dreams are all for individuals, not for helping everyone escape the slums. Should the book talk about widening people's circle of compassion to include not just one's family or one's race, religion or nationality but perhaps also to include non-human animals (Jacobs, 2004)? This relates to the issue of meritocracy, which says that in a just society, people should have the opportunity to compete on a level playing field for success; for example, people should not be given prestigious, well-paid positions because of who their families are – people should earn their success via hard work and intelligence. Minni and Sanjay are hoping to do this.

Sandel (2020) and Lie et al. (2022) highlighted the negative side of meritocracy, that it provides, however unwittingly, a rationale for assigning large portions of society, such as those who do manual labour (e.g. cleaning), to low status and low income lives on the grounds that they had their opportunity to compete, but they did not work hard enough and/or were not smart enough to achieve the success that others rightfully earned. In other words, those with higher income and higher status too often absolve themselves from any obligation to care for their lower-income, lower-status fellow humans by telling themselves and others that they deserve their favourable positions in society.

Actually, much in the book *Thirst* does encourage a wide circle of compassion. For example, Minni's family are Hindus, but both the children's best friends are Muslims. The children's parents collaborate as well, such as when Sanjay and his best friend Amit need to leave Mumbai because they are in danger from the Water Mafia. The two boys go to a farm where one of Amit's uncles works. Another example of collaboration across cultures can be seen in the computer class that Minni attends. Shanti is a neighbourhood storyteller, advisor, historian and general spreader of harmony and wisdom. She comes to Minni's computer class to help the students interact with the teacher, Priya, who, in another example of kindness, has just arrived from the United States to teach IT to girls in the slums. Though Priya is full of good intentions, she is unfamiliar with some aspects of local culture and is grateful for Shanti's guidance.

Indeed, Shanti may offer the wider perspective that Minni, as a child, lacks. She has a broad view of the community and works for the overall betterment of everyone. For example, when Shanti tells the story of how Mumbai (then Bombay) was 'given' by Portugal to England, one of the children asks whether the people of Mumbai were also 'given'. 'Were we like cattle, of no importance back then?' Sanjay, Minni's fifteen-year-old brother, replies, 'We aren't exactly important now either, which is why we don't get water. And get questioned before we enter fancy shops. As if we might steal.' '"You are bringing up an important point," Shanti says, using her best teacher voice.'

Shanti's husband and daughter had died in an accident, after which she studied to become a teacher as well as volunteering at the local community centre. She feels that 'all the children in the neighbourhood are now her children'. She clearly seeks to be a leader in the community's march towards progress. When Minni comes to her for advice about all the problems she is facing in the wake of her mother and brother leaving Mumbai, Shanti assures her, 'You're not alone, dear Minni. And you're very brave.' When Minni and Faiza decide to expose Pinky's father as a leader of the Water Mafia, they come to Shanti asking whether regular people can defeat the powerful. Shanti replies, 'Why not? Remember the story of the foolish lion?' This is a story from Panchatantra, from about 200 BC, in which a clever rabbit defeats a greedy lion.

The above notwithstanding, it should be noted that, in *Thirst*, all is not harmony in the slum. For example, sometimes disputes in the queues for water degenerate into fights. However, in the book, such conflicts are attributed not to any innate selfishness of humans or of poor, lower caste people in particular but to the shortage of water, a result of the existing power structure. In general, people in the slum are depicted as helping each other and socializing together, whereas such a spirit of community does not seem to exist where Pinky and her family live, and Pinky envies the social connections that Minni, Faiza and their families enjoy. Indeed, research suggests that community ties positively correlate with enhanced longevity.

Narratives

Stibbe's ninth category concerns narratives. As Newman (2003) noted in a paper titled 'The power of stories over statistics', stories can often communicate better than data. For example, in a newspaper article on the problems of people living with worms in their bodies, Kristof (2023) began with statistics about the problems, such as that it affects 1.5 billion people, and described the costs involved to treat people. Next, however, Kristof told stories, accompanied by photos, of two people who had suffered greatly due to worms. Perhaps Bajaj, the author of *Thirst*, has been able to reach many more people and more effectively on the issue of water by talking about the issue as a story. In the Author's Note at the end of the book, Bajaj does include statistics, along with websites where readers can learn more and take action to help people such as Minni.

Conclusion

This chapter has used a children's book written for preteens to illustrate concepts from ecolinguistics that are applicable to children's literature and all other forms of media. These concepts seek to encourage critical thinking about human actions towards the environment and towards our fellow humans. Of course, thinking without participation in the real world may have little impact. The United Nations Environmental Programme (1976) proposed six environmental education objectives culminating in objective #6, participation. However, Jacobs and Goatly (2000) found very little participation in the many discussions of the environment in seventeen ESL textbooks. Let us hope that future students and teachers can do better.

References

Ahlquist, S. (2023). Integrating children's fiction and Storyline in the second language classroom. *Education Inquiry*, *14*(1), 105–124. https://doi.org/10.1080/20004508.2021.1965287

Bajaj, V. (2022). *Thirst*. Penguin.

Benchley, P. (1974). *Jaws*. Doubleday.

Chau, M., Zhu, C., Jacobs, G. M., Delante, N., Asmi, A., Ng, S., et al. (2022). Ecolinguistics for and beyond the sustainable development goals. *Journal of World Languages*, 8(2), 323–345. https://doi.org/10.1515/jwl-2021-0027

Cho, K. S., & Krashen, S. (2016). What does it take to develop a long-term pleasure reading habit? *Turkish Online Journal of English Language Teaching*, *1*(1), 1–9.

Halliday, M. A. K. (1990). New ways of analyzing meaning: A challenge to applied linguistics. *Journal of Applied Linguistics*, 6, 7–36.

Harvard Medical School. (2020, April 15). *Becoming a vegetarian*. Retrieved from: https://www.health.harvard.edu/staying-healthy/becoming-a-vegetarian

Hintz, C., & Tribunella, E. L. (2019). *Reading children's literature: A critical introduction*. Broadview Press.

International Ecolinguistics Association. (2023). *About*. https://www.ecolinguistics-association.org

Ismail, H. M. (2023). Children's literature: The significance and other impacts. *Theory and Practice in Language Studies*, *13*(3), 593–598. https://doi.org/10.17507/tpls.1303.07

Jacobs, G. M. (2004). Extending the circle of compassion to include nonhuman animals: The case of the use of who in dictionaries, works on grammar, and publication manuals, and by newspapers and news agencies. *Language & Ecology*. https://citeseerx.ist.psu.edu/document?repid=rep1&type=pdf&doi=c2e2e031aa0dfec51b8aaf23fb314a5ba614234f

Jacobs, G. M., & Goatly, A. (2000). Ecological issues in ELT coursebooks. *ELT Journals*, *54*, 256–264. https://files.eric.ed.gov/fulltext/ED574083.pdf

Koenig, K. (2023, May 19). Jeff Bezos' new yacht is finally ready to set sail. *New York Times*. https://www.nytimes.com/2023/05/19/style/bezos-yacht-koru.html?

Kristof, N. (2023, May 20). How to change a life for 50 cents. *New York Times*. https://www.nytimes.com/2023/05/20/opinion/sierra-leone-deworming.html

Le Guin, U. K. (1992). *Language of the night: Essays on fantasy and science fiction*. Harper Collins.

Leland, C. H., Lewison, M., & Harste, J. C. (2017). *Teaching children's literature: It's critical!* Routledge.

Lie, A., Chau, M. H., Jacobs, G. M., Zhu, C., & Winarlim, H. S. (2022). Exploring the role of English in meritocracy in multilingual Indonesia: Teacher beliefs, language use, and Global Englishes. *Journal of Multilingual and Multicultural Development*, *45*(10), 4200–4217. https://doi.org/10.1080/01434632.2022.2151601

Lim, S. C. (2009). *Granny* (revised ed.), (R. Foo, Illus.). EPB Pan Pacific.

Lim, S., Reidak, J., Chau, M. H., Zhu, C., Guo, Q., Brooks, T. A., Roe, J., & Jacobs, G. M. (2023). *Cooperative learning and the SDGs*. Peachey Publications. https://payhip.com/b/obOwr

Ma, C., & Stibbe, A. (2021). The search for new stories to live by: A summary of ten ecolinguistics lectures delivered by Arran Stibbe. *Journal of World Languages*, *8*(1), 164–187. https://www.degruyter.com/document/doi/10.1515/jwl-2021-0031/html

Martin, J. R. (2004). Positive discourse analysis: Solidarity and change. *Revista Canaria de Estudios Ingleses*, *49*, 179–202.

Newell, P., Srivastava, S., Naess, L. O., Torres Contreras, G. A., & Price, R. (2021). Toward transformative climate justice: An emerging research agenda. *Wiley Interdisciplinary Reviews: Climate Change*, *12*(6), e733. https://doi.org/10.1002/wcc.733

Newman, T. B. (2003). The power of stories over statistics. *The BMJ*, *327*(7429), 1424–1427. https://doi.org/10.1136/bmj.327.7429.1424

Penz, H., & Fill, A. (2022). Ecolinguistics: History, today, and tomorrow. *Journal of World Languages*, *8*(2), 232–253.

Sandel, M. J. (2020). *The tyranny of merit: What's become of the common good*? Penguin.

Stibbe, A. (2021). *Ecolinguistics: Language, ecology and the stories we live by* (2nd ed.). Routledge.

UNESCO–UNEP. (1976). The Belgrade Charter: A global framework for environmental education. *Connect*, *1*(1), 1–9.

UNICEF. (2019). *1 in 3 people globally do not have access to safe drinking water*. https://www.who.int/news/item/18-06-2019-1-in-3-people-globally-do-not-have-access-to-safe-drinking-water-unicef-who

United Nations. (2023). *The 17 goals*. https://sdgs.un.org/goals.

Webb, S., & Macalister, J. (2013). Is text written for children useful for L2 extensive reading? *TESOL Quarterly*, *47*(2), 300–322. https://doi.org/10.1002/tesq.70

Williams, C. (Director). (2022). *The sea beast* [Film]. Netflix.

Zhang, Y., Hawk, S. T., Opree, S. J., De Vries, D. A., & Branje, S. (2020). "Me", "we", and materialism: Associations between contingent self-worth and materialistic values across cultures. *The Journal of Psychology*, *154*(5), 386–410. https://doi.org/10.1080/00223980.2020.1759496

5

Integrating ecolinguistic perspectives of Outstanding Universal Values of *subak* in English learning

A mental model analysis

I Gusti Agung Sri Rwa Jayantini, Sang Putu Kaler Surata and Ronald Umbas

Introduction

In Bali, *subak* – the traditional irrigation system – is a paramount cultural asset that has high potential to create environmental and cultural awareness. Officially designated by UNESCO as a World Heritage Site with Outstanding Universal Value (OUV), *subak* not only plays a vital role in agriculture but also has a strong connection to the life, rituals and existence of the Balinese people, as its values share manifestations of the philosophy of *Tri Hita Karana* (the three causes for well-being and prosperity). Balinese life with *subak* can teach the world about perfect order, which implies honouring all three relationships to make a balanced life, that is, the relationship between man and God known as *Parahyangan*, the relationship between humans and nature known as *Palemahan*, and the relationship humans have with each other known as *Pawongan* (Asmiwyati et al., 2015; Niswatin & Mahdalena, 2016; Sriyanti & Sriartha, 2019).

Subak is well-known as an inspiration for its integrity and authenticity. The UNESCO website (https://whc.unesco.org/en/list/1194/) profiles *subak* in terms of these two significant aspects. Integrity is reflected in the way the land is shaped by farming, which is still vibrant and resilient because of the well-managed *subak* system (Adnyana, 2016; Jayantini et al., 2023). The management system needs to consistently support traditional farming methods to continue to give farmers the help they need to stay on their land (Puspawati et al., 2018). The way people and nature interact is the core value in *subak*. With regard to authenticity, the *subak* system is known for its distinctive aspects of Balinese culture (Febrianto et al., 2021; Roth, 2011). The principles of *subak*, as a traditional communal farmer system, guide irrigation water management and agricultural planting techniques in accordance with *Tri Hita Karana* ideology (Sriartha et al., 2015).

The results of a significant number of research studies demonstrate that *subak* provides a variety of universal benefits related to local knowledge (Sriartha et al., 2017; Surata, 2013; Surata et al., 2014). In the context of learning in Bali, finding out more about the potential utilization of local values related to the environment in the education world is strongly needed because the relationship between humans and the environment cannot be separated from the traditional, balanced life of the Balinese people (Jayantini, 2022). As Bali has become a world tourist destination where English is mastered by the younger generation to obtain jobs in tourism, this chapter attempts to map the mental models of the younger generation onto their *subak* and natural surroundings so that the results can be used to encourage a new ethical dimension to the English language in English-language learning in secondary education. Overall, this chapter identifies the Balinese younger generation's mental model that reflects ecolinguistic knowledge through how they express ideas about their environment.

In this chapter, we will report from an innovative study in which over 150 students and teachers from secondary schools in Bali drew pictures of *subak* and were then interviewed about their drawings. These pictures were subsequently analysed in terms of mental model theory (Johnson-Laird, 1983), according to criteria previously established by Subramaniam (2019). The results shed light on how younger people and educators conceptualize their cultural heritage, and on how ecolinguistic awareness might be better integrated into English-language learning.

The Outstanding Universal Value of *subak*

As a cultural landscape, *subak* reflects the life of Balinese Hindu community with its traditional agricultural and water management practices (Roth, 2011; Salamanca et al., 2015). The values of *subak* incorporate ideas for a sustainable way of life that has been developed for over two millennia. The preserved values go beyond just being a traditional farming method. Inspired by the concept that 'the rice cycle is the cycle of life' (Reynolds, 2013), the values of *subak* depict human existence and the life of other living things as a cycle that starts with birth, continues with life and ends with death. The cycle is a true model of sustainable living. Water that flows through volcanic mountains is the primary supply to irrigate the paddy fields (Reynolds, 2013). The water is utilized to grow many crops of rice, which help the soil retain nitrogen and other nutrients (Surata et al., 2024).

Subak has become the subject of study in various disciplines, one of which is a particular type of pedagogical approach known as heutagogy, a student-centred approach to learning in which students identify problems and work towards solutions with a high degree of autonomy. This approach is recommended for its flexibility and adaptability in the context of technological advances (Arini & Sudatha, 2023). Moreover, the values of *subak* as agricultural cultural heritage in Bali have ample potential in the context of education (Risna et al., 2022; Parwata et al., 2018). The local wisdom of *Tri Hita Karana* has global relevance, as it supports the pillars of lifelong learning from UNESCO (Arjaya & Diarta, 2018; Mantaka et al., 2017). Integrating Balinese local wisdom can also improve learning outcomes (Lestari et al., 2017; Sriyanti & Sriartha, 2019). Previous research suggests the inclusion of environmental vocabulary, which are phrases that students can use to explain agriculture, the environment and the breathtaking views of the natural world. There is a positive attitude among teachers and students regarding the potential for creating the recontextualization of the *subak*'s OUVs in English instruction (Surata et al., 2014; Jayantini et al., 2023).

Conceptualizing mental models

A mental model of an individual is a private, subjective representation of their working knowledge that shows how they take in and reject information to create a worldview, and how they act in certain circumstances (Johnson-Laird, 1983; Greca & Moreira, 2000; Jones et al., 2011). The conceptualization of individuals' awareness of the environment through learning – especially from direct experience – plays a crucial role in shaping people's mental models (Surata et al., 2024; Wuellner et al., 2017). According to Shepardson et al. (2007), in order for individuals to completely comprehend and regulate their behaviour in the environment, they must first conceive of the relevant phenomena using a mental model. Mental models can shed light on how students behave in the environment going forward. In the Indonesian context, applying mental models in order to develop appropriate educational policies is considered beneficial so that the richness in local biodiversity and cultural knowledge can be included, and the younger generation can draw inspiration from their ancestral history to construct the future (Surata et al., 2024).

Drawings have been utilized in numerous studies to gather data regarding the involved participants' views and ideas across a range of themes and ages. The findings have demonstrated that drawing is a useful technique to reveal how people view particular locations, procedures, incidents and their depictions of reality, as well as their attitudes and feelings (Surata et al., 2014). Morón-Monge et al. (2021) propose acquisition strategies in education that can be designed using drawing as a foundation. Drawing can be used with different learning objectives and instructional methods, giving the instructor new tools for creative and efficient teaching techniques (Johnson-Laird, 1983). Teachers are often drawn to the idea of using students' drawings to understand their mental images in relation to their traditional values and practices.

An ecolinguistic perspective in English-language learning

Considering the importance of recognizing the views of teachers and students in utilizing ecolinguistic knowledge associated with *subak*, the discussion is expected to recommend the inclusion of environmental and cultural awareness by promoting Balinese local values in English teaching and learning. To achieve this goal, our study involves mapping through drawing in order to classify the relevant mental models and identify environment-bound vocabulary. Previous research has discussed green contexts for learning, particularly language learning, and how ecological issues can be part of teaching materials and textbooks at various levels of education. Such studies include, for example, contexts of English-language learning (Ekasiwi & Bram, 2023; Mliless & Larouz, 2018; Zahoor & Janjua, 2020), teaching Indonesian as a foreign language (Pratiwi et al., 2022) and Indonesian textbooks (Ekasiwi & Bram, 2023). There is a consensus that ecolinguistic integration in language learning is important, whether it is a local, national or foreign language.

That said, the ecolinguistic integration of *subak* into English teaching materials and activities in Bali has been rarely explored. Arguably, linking local values as a legacy of our ancestors is significantly important to provide additional value to learning. To realize this intention, Jayantini et al. (2023) conducted a qualitative study with teachers and students in elementary and secondary school to determine the potential of drawing the Outstanding Universal Value of *subak*. The authors found that there was a positive attitude among teachers and students regarding the integration of traditional values. The current study builds on the findings of Jayantini et al. (2023) to shed further light on this exciting educational initiative.

Methods

This research utilized a mental model analysis drawing on data obtained through four main activities. There were four steps: (1) the students and teachers drew a picture representing their natural surroundings including the rice fields, (2) the students listed the English vocabulary for objects in their drawings, while the teachers made a short description of the picture consisting of one or two paragraphs, (3) the students and teachers were interviewed about their opinions on the importance of *subak* in creating a green learning environment in English classrooms, and (4) all the gathered data were analysed in the form of drawings supported by the information collected from the interviews. All of these steps were conducted to map mental models showing the ecocultural knowledge and images that have the potential to be utilized as part of more enlightened English teaching materials.

Research site and participants

The research site was the *subak* system of 19,500 ha of rice terraces in Bali and twelve schools, comprising eleven public schools and one private school, situated near this landscape. The participants were 119 students aged 13–15 years and 10 teachers, who all gave their informed consent to participate in the study. UNESCO's World Heritage Site with its 19,500 ha in Bali consists of five rice terraces and their water temples, reflecting the philosophical concept of *Tri Hita Karana*, which brings together the world of spirit, the world of human and nature. The 129 participants who were involved voluntarily participated after signing consent forms giving their permission that the collected data could be primarily used in research activities. All participants came from twelve secondary schools around the scenic natural landscape that is a UNESCO precious heritage site in Bali.

Data collection

The research involved a triangulation of data: namely (1) participants' drawings, (2) written responses mentioning the objects that the participants included in their drawing, and (3) semi-structured interviews (conducted following guidelines in Ruslin et al., 2022). The students were asked to write English words related to their drawings, and teachers were asked to write a description in one or two paragraphs about the drawings on the same paper, on the back of the pictures. The words and the short descriptive writing had to be in English.

Drawings and written responses were collected to explore the perspectives of students and teachers on their natural surroundings. The arrangement was realized in two activities. In the first lesson (forty-five minutes), the participants received necessary information about the purpose of the research and the nature of the activities. Each participant was given a blank A4 sheet. They were asked to draw scenery in *subak* areas. The participants were free to make black-and-white or colour drawings. Additionally, they were told that the researchers' primary interest was the objects in the students' drawings, regardless of how well they were drawn.

Data analysis

Data were analysed using procedures modified from Subramaniam (2019). The initial step was reviewing all the drawings until the content was clearly understood. Next, four categories of mental model patterns were created by identifying and evaluating similar content in all drawings. Further review involved considering the objects in the drawings and the vocabulary listed by the participants, which were further classified by the researchers. All schoolchildren were able to create vocabulary lists in English without assistance from their teachers. At this stage, each participant's drawing was then analysed along with each student's written document to identify, classify and categorize the codes for mental models. To better understand how individuals and groups perceive the use of traditional values in teaching materials, interviews were also conducted.

Mental Model Categories

Following mental model theory (Johnson-Laird, 1983) in accordance with criteria previously established by Subramaniam (2019), four types of mental models were proposed on the basis of the beliefs, assumptions and conceptual frameworks that appear to have guided the drawings by participants. Of the 129 drawings, only 122 were analysed in consideration of their suitability for the four categories. It was found that Mental Model Category 1 (MMC1) included 105 pictures (86 per cent); Mental Model Category 2 (MMC2) was represented by twelve pictures (10 per cent); Mental Model Category 3 (MMC3) was found in two pictures (2 per cent); and Mental Model Category 4 (MMC4) involved three pictures (2 per cent). For the list of vocabulary, the students and teachers were able to use words and phrases related to scenic rice fields. Teachers were also able to write a descriptive paragraph to explain what their picture was about. They could describe the scenic environment by employing words and phrases which were used to express their positive feelings towards nature. Students' and teachers' positive attitudes towards the inclusion of environmental topics, including *subak*, were demonstrated during the interviews. All participants thought that environmental issues were crucial for the younger generation to increase their awareness. As a promising action, teachers even planned to choose topics related to *subak*, rice fields, farming and other environmental themes as topics in speaking, writing, reading or listening activities for students. The four mental model categories help to visualize the relationships between different elements with the concepts that give rise to them, as shown in Table 5.1.

Category 1: Natural landscape with/without buildings

Mental Model Category 1 (MMC1) shows the basic reflections in participants' minds of what they find in their surroundings. They drew living and non-living things – biotic and abiotic components – as well as buildings. This demonstrates the connection of nature and buildings for humans to shelter. The participants' drawings did not include humans. In this category, the students and teachers created pictures of non-living things like mountains, the sun, the sky and rocks. In addition, the students drew 'typical' drawings of scenery showing objects like rice terraces, rivers, roads and houses. They also drew plants like grass and trees, the sun and clouds, as well as birds.

Table 5.1 Category of Mental Models and Their Dominant Characteristics

Category of Mental Model	Dominant Characteristic of Mental Model	Reflected Ecolinguistic Perspective	List of English Vocabulary
Mental Model Category 1 (MMC1)	Natural landscape with/without buildings	Biotic and abiotic elements are found in participants' drawings, including the following items: • Animals and plants living with buildings • Animals and plants living without buildings • Plants with buildings • Plants without buildings	Words used to talk about scenic rice fields: mountain, river, stone, plant, rice field, sun, cloud, house, grass, flower, water etc.
Mental Model Category 2 (MMC2)	Natural resources that support humans	Biotic and abiotic elements are found in participants' drawings including the following items: • Humans with natural resources and buildings • Humans with natural resources without buildings	Same as words in MMC1 with the addition of phrases like rice terraces, coconut tree, etc.
Mental Model Category 3 (MMC3)	Manifestation *of Tri Hita Karana*	Biotic and abiotic elements are found in participants' drawing including the following items: • *Tri Hita Karana* (temple, human and environment) with buildings • *Tri Hita Karana* (temple, human and environment) without buildings	Same as MMC1 and MMC2 with the addition of the word 'temple' as one aspect that shows the relationship between God and human
Mental Model Category 4 (MMC4)	*Tri Hita Karana* related to biocultural diversity	*Tri Hita Karana* (temple, human and environment) with rice terraces, water irrigation and threats of the natural world like pests and diseases	Same as MMC1, MMC2 and MMC3 with the addition of 'water irrigation' as the core activity of rice farming

Figure 5.1 (a, b) reflects examples of MMC1 that have only birds as 'animals' and plants found with building elements, or houses. This shows that the participants understand *subak* as a natural landscape captured through rice terraces, as shown in Figure 5.1 (a). The picture shows a beautiful rice terrace, but the paddy has not grown yet. From an ecolinguistic perspective, when

(a) (b)

Figure 5.1 Category 1: Natural landscape with/without buildings.

drawing a picture of the terraces, they were able to mention basic words to indicate the biotic and abiotic elements present in the rice field environment. Figure 5.1 (b) appears to be a very common drawing of scenery with three mountains and the rice paddies under the sun. Plants and trees existing in nature are the core living things captured to represent the picture of the natural landscape.

These two pictures were chosen to show the mapping of the neighbourhood of natural landscapes as the physical components of *subak*. A house is seen in Figure 5.1 (b), which is in contrast to Figure 5.1 (a), where it could be said to be 'only rice fields without buildings established nearby'. Figure 5.1 (a) shows a paddy field without rice, reflecting an early stage in the farming process. There is a period before planting rice, a period of nurturing rice and a period when the harvest arrives to create a different atmosphere in the neighbourhood. In addition, terraced soil, water and mountains as abiotic elements support a natural ecosystem and the life cycle of farmers who maintain the rice fields. Figure 5.1 (b) shows a more complex ecosystem in the rice fields. Figure 5.1 (a) does not show the growing crops, but Figure 5.1 (b) has been filled with rice and shows the clear flow of the river. Both images show common landscape features that children and even adults draw in showing natural landscapes. The most common elements are mountains, roads and clouds as abiotic components, with trees and birds as biotic components.

Category 2: Natural resources that support humans

The second Mental Model Category (MMC2) presents *subak* as having natural resources and humans that support the system. Figure 5.2 shows nature

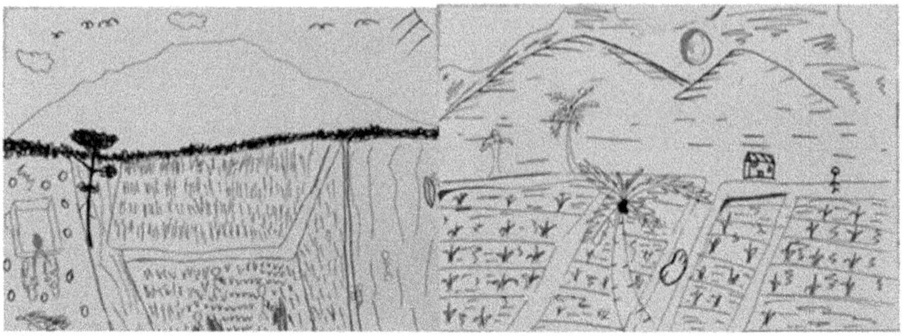

(a) (b)

Figure 5.2 Category 2: Natural resources that support humans.

supporting agriculture with the help of living creatures providing convenience and benefits to humans and their agricultural activities. This second mental model category is classified by considering several characteristics, namely the presence of types of living things such as animals and plants, that appear to be highlighted as one of the elements of life. This ecosystem filled with biotic and abiotic elements emerges similarly to a typical scenic picture but with an emphasis on how the biotic elements in the picture contribute to human life.

Figure 5.2 (a) shows a rural ecosystem in a mountainous area. The mountain is complemented by rice fields and a flowing river, illustrating humans as part of an ecosystem with its biotic and abiotic components. This picture demonstrates the role of humans in cultivating nature, and that nature provides blessings for humans. These blessings are in the form of nature containing other living creatures that are sources of human life, particularly rice as a staple food. It is interesting to see some farmers working in the rice fields, depicted in the picture ploughing the fields, taking care of their rice plants and cultivating the land before the rice seeds are planted.

Meanwhile, Figure 5.2 (b) has as its main focus beautiful scenery where rice fields are shown with a small river that irrigates each area. This image is included in Category 2, which shows that the universe supports human life. Unlike Category 1, which comprises images of only landscapes, Category 2 shows people in the rice fields, which, although they appear small in the picture, carries the message that nature is part of human life. The environment with rice fields and houses is completed with humans blessed with abundant nature to be cultivated. Figure 5.2 (b) shows the relationship between humans and nature, which in the picture is depicted in a rice field environment. It is presented by fairly large and shady coconut trees, and water as irrigation that makes the soil fertile. This was explained by one teacher in her description. This teacher, whose writing was more beautifully written than the others, employed commonly used English words in order to provide a vivid description of the impressive scenery of a rice field.

> The picture describes about beautiful rice field. We can see river beside the rice field. Farmers always use the water in the river to irrigate the fields. There is a coconut tree beside the river. The coconut tree is very lush. It gets nutritions from the river and land. The picture is beautiful and green. (Tcr-Mw1, female)

Category 3: Manifestation of *Tri Hita Karana*

Mental Model Category 3 (MMC3) shows the components of *Tri Hita Karana* delineated in the pictures, illustrating a way of life practiced in Bali, Indonesia. A well-maintained relationship between God, people and the environment underlies the beliefs and spiritual practices of the people of Bali. This is evident in how they live their daily lives, including the arrangement of their living spaces. Spatial planning with *Tri Hita Karana* includes building temples for worship, taking care of nature and considering human activities. Pictures with temples, people and nature are categorized in MMC3, which encompasses temples and shrines in the natural environment.

Figure 5.3 (a, b) represents the idea of *Tri Hita Karana*, which describes how Karana temples are integral to Balinese people's lives and part of agricultural activities. The two pictures show similar views. Humans and temple were drawn to illustrate the connection with God, which is a core value for Hindu devotees. Figure 5.3 (a) shows two people on their boat, sailing in search of

(a) (b)

Figure 5.3 Category 3: *Tri Hita Karana* manifestation.

fish. Meanwhile, Figure 5.3 (b) also demonstrates life along the lake shore, highlighting the existence of fishermen. The picture of agricultural life in Bali, based on the concept of three connections in human life, is clearly visible in both pictures. The presence of the temple, even though it appears very small in the picture, is an important part of both illustrations, reinforcing the third model category.

Another interesting aspect interpreted from Figure 5.3 (a, b) is what nature gives humans. Apart from agriculture, life also provides opportunities for humans to become fishermen at the lake, the main source of water for agriculture in Bali. Living in harmony with nature is shown in both images. The view is beautiful because of the meeting of mountains and lakes supported by green nature, which gives the impression of a peaceful and beautiful life. The student who drew Figure 5.3 (a) stated this clearly during the interview.

> A very beautiful scenery around Tamblingan Lake [one of the four lakes of the island of Bali] can be seen. There are several mountains and still green with different plants. Fishermen can still catch fish easily. There is a temple on the edge of the lake. (Std-Kt4, female)

Category 4: Biocultural diversity reflecting *Tri Hita Karana*

A biocultural landscape mental model is linked to how people see landscapes that help life – how people and the environment work together – as part of *Tri Hita Karana*, reflecting rice fields and water systems. Based on these concepts, the two selected pictures were matched with Mental Model Category 4 (MMC4). They show temples, people and things around them, as well as rice fields on both sides of the road. A biocultural landscape mental model is a picture of places where people and nature interact, like rice fields and irrigation systems. It also includes the idea of *Tri Hita Karana*: a way to live in harmony with nature.

Based on the characteristics of MMC4, two pictures are presented as examples of this model category. Temples, people, animals, plants and things around them, water irrigation, and rice fields on both sides of the road are captured. In Figure 5.4 (a), it can be seen that biodiversity exists as part of the rice field landscape, which indicates the value of respecting natural surroundings to achieve a harmonious life by creating a relationship between the three components of this philosophy to create a balanced life: *Parahyangan*, *Palemahan* and *Pawongan*. Figure 5.4 (b) shows the complexity of biodiversity. It reflects the values of a balanced life shown in the existence of temple, human and environment. The rice fields

(a) (b)

Figure 5.4 Category 4: Biocultural diversity reflecting *Tri Hita Karana.*

are clearly visible in the picture, with the presence of plantation areas with a mountain and river background that represents farming activity. Figure 5.4 (a, b) shares the completeness of the *Tri Hita Karana* elements, featuring temples, humans and nature in the pictures. The completeness of the components in farming, including plant pests and diseases and water canals as the source of irrigation, demonstrates the biocultural diversity.

In this category, there was another picture having the same characteristics (the total number of drawings in this category was three). The other picture, which is not included in any of the figures, offered an interesting insight into the teacher's critical thinking. The teacher captured chemical pesticides that are widely used because they have the advantage of being applied easily, and the results can be felt in a relatively short time. Chemical pesticides are sprayed in the area of orange trees and must be managed properly. This phenomenon reflects a biodiversity challenge, where humans strive to preserve their fundamental life principles while overcoming problems like pests that threaten their crop. However, humans must be wise and learn to reduce the use of chemicals as much as possible for sustainable living.

Incorporating *subak* in English-learning activities

From the drawing activities, listing vocabulary by students and writing a short paragraph by teachers, an ecolinguistic perspective in English learning has actually been achieved. All these activities are beneficial for enhancing students' production skills, like speaking and writing, by utilizing drawings that the students can then describe orally or in written form. Teachers can use their writing as examples to help students be able to describe nature with a focus on the rice fields. As a continuation of these activities, teachers could ask students to make sentences from the words they listed. Exploring more learning materials on the theme of harmonizing Balinese life and the *subak* irrigation system, such as short articles from books or the internet for students to read, summarize, present and discuss, could become an important step in teaching preparation.

Furthermore, to focus on the enrichment of vocabulary, many exercises could be adopted, such as finding synonyms and antonyms, and mentioning other parts of speech that relate to the nouns given by students when listing the vocabulary. To incorporate *subak* in English-learning activities, teachers need to give students a literal understanding. The explanation could include how *subak* shows how human beings cooperate in the system that manages irrigation traditionally. The OUV could be discussed when students begin to learn about text types as being descriptive, explanatory or argumentative. Themes could be developed into fascinating learning materials that discuss, for example, the role of *subak* temples in the landscape, also known as *Ulun Carik* temples or *Bedugul* temples, and farmers' activities and rituals that are carried out to maintain tradition, nature and the worship of *Dewi Sri*.

Thus, the mapping of mental models could be utilized to maintain local wisdom within the general mission to improve English-language skills. It should also aim at increasing concern for the environment and preserving local wisdom (Niswatin & Mahdalena, 2016; Sriartha et al., 2017; Wirahyuni et al., 2021). The values that encapsulate the *Tri Hita Karana* philosophy (Ronny & Mahendra, 2013) must be preserved by the younger generation, even as they learn English and are exposed to other values in the international community. The life of Balinese people who respect the Outstanding Universal Value of *subak* can stand as an example to the world about harmony and sustainability (Lansing, 2012). This value implies respecting the three relationships for a balanced life, namely the relationship between humans and God, known as *Parahyangan*, the relationship between humans and nature, known as *Palemahan*, and the relationship between humans and each other, known as *Pawongan* (Asmiwyati et al., 2015 ; Niswatin & Mahdalena, 2016; Risna et al., 2022).

Conclusion

On the basis of student and teacher drawings, four mental model categories were created that contained the characteristics of (1) natural landscape with/without buildings; (2) natural resources that support humans; (3) the manifestation of *Tri Hita Karana*; and (4) biocultural diversity reflecting *Tri Hita Karana*. However, despite the four mental model categories that emerged from the analysis and captured how the secondary school students and teachers were able to reflect on their understanding of their ancestors' legacy, the findings in general show that the students have not recognized that the existence of *subak* is not just about rice fields and beautiful scenery. Through a drawing activity to map the mental model combined with listing vocabulary and writing a descriptive paragraph, an ecolinguistic approach by exploring *subak* with its OUV has possibly been applied. To develop students' major language skills within an exploration of *subak* and environment, English-learning activities could become a fun activity through teacher creativity. Teachers could incorporate the topic of the Balinese irrigation system while exploring students' speaking skills. With the vocabulary about nature that the students already know, they can be directed to speaking and writing as productive skills. English teachers can utilize *subak* to enhance students' environmental awareness by including *subak*, the Balinese farming system, the lives of Balinese farmers and other related topics when designing their teaching activities.

The application of environmental awareness-based English-learning activities at the secondary education level is a strategy that includes the ecolinguistic approach. This encourages students to engage with environmental topics and uses language learning to promote ecological awareness and responsibility. By incorporating *subak* with its OUV as the theme for drawing and listing vocabulary, which is further developed into speaking, writing, reading and listening activities, discussions about sustainability and environmental ethics can be integrated into English lessons. Through these engaging activities, learners not only develop their language skills but also become more conscious of their role in supporting ecological balance, making language learning a tool for fostering a more sustainable world.

References

Adnyana, B. (2016). Subak Sebagai Media Pembelajaran Biologi Berbasis Kearifan Lokal. *Prosiding Seminar Nasional MIPA*, 142–147.

Arini, N. M. A., & Sudatha, I. G. W. (2023). Bahan Ajar Muatan IPS Berpendekatan Heutagogy Berbasis Kearifan Lokal Bali Sistem Subak. *Jurnal Ilmiah Pendidikan Profesi Guru*, *5*(3), 623–635. https://doi.org/10.23887/jippg.v5i3.57798

Arjaya, I.B.A & Diarta, I.M. (2018). Penerapan Model Pembelajaran Berbasis Proyek (PJBL) terhadap Hasil Video Digital Story Telling Landskap Budaya Subak. Seminar Nasional Pendidikan Biologi. https://jurnalfkip.unram.ac.id/index.php/SemnasBIO/article/view/708

Asmiwyati, I. G. A. A. R., Mahendra, M. S., Arifin, N. H. S., & Ichinose, T. (2015). Recognizing indigenous knowledge on agricultural landscape in Bali for micro climate and environment control. *Procedia Environmental Sciences, 28*(SustaiN 2014), 623–629. https://doi.org/10.1016/j.proenv.2015.07.073

Ekasiwi, A. N., & Bram, B. (2023). Ecolinguistic topics in Indonesian English textbooks. *Interference: Journal of Language, Literature, and Linguistics, 4*(1), 20. https://doi.org/10.26858/interference.v4i1.42935

Febrianto, M., Marsiti, C. I. R., & Damiati, . (2021). Identifikasi Potensi Subak Sambangan Sebagai Daya Tarik Ekowisata Di Desa Sambangan. *Jurnal BOSAPARIS: Pendidikan Kesejahteraan Keluarga, 12*(3), 118–127. https://doi.org/10.23887/jppkk.v12i3.40526

Greca, I. M., & Moreira, M. A. (2000). Mental models, conceptual models, and modeling. *International Journal of Science Education, 22*, 1–11.

Jayantini, I. G. A. S. R., & Hikmaharyanti, I. P. D (2022). Eco-cultural approach through local wisdom identification in designing teaching materials for English department students. *International Conference on Language, Culture, and Communication (ICLCC 2022)*, 35–41. https://eproceeding.warmadewa.ac.id/index.php/iclcc/article/view/14%0Ahttps://eproceeding.warmadewa.ac.id/index.php/iclcc/article/download/14/11

Jayantini, I. G. A. S. R., & Surata, K. (2023). *Proceedings of the International Conference on Language, Education, and Social Science (ICLESS 2022)*. Atlantis Press SARL. https://doi.org/10.2991/978-2-494069-15-2

Jayantini, I. G. A. S. R., Surata, S. P.K., & Yuniti, I. G. A. D. (2023). Utilizing outstanding universal values of subak to design a green textbook for English learning (A preliminary study). *International Conference on Language, Education, and Social Science (ICLESS 2022)*. Atlantis Press SARL. https://doi.org/10.2991/978-2-494069-15-2

Johnson-Laird, P. N. (1983). *Mental models : Towards a cognitive science of language, inference, and consciousness*. Harvard University Press.

Jones, N. A., Ross, H., Lynam, T., Perez, P., & Leitch, A. (2011). Mental models: An interdisciplinary synthesis of theory and methods. *Ecology and Society, 16*(1). https://doi.org/10.5751/ES-03802-160146

Lansing, J. Stephen. (2012). *Perfect Order*. Princeton University Press.

Lestari, N. K. T., Kristiantari, M. R., & Ganing, N. N. (2017). Pengaruh Model Pembelajaran Talking Stick Berbantuan Lagu Daerah Terhadap Hasil Belajar IPS. *International Journal of Elementary Education, 1*(4), 290. https://doi.org/10.23887/ijee.v1i4.12960

Mantaka, I. N., Sendratari, L. P., & Margi, K. (2017). Pengintegrasian Kearifan Lokal Subak Abian Catu Desa Sambirenteng Buleleng Bali Sebagai Sumber Belajar Ips Di

Smp. *Jurnal Pendidikan IPS Indonesia, 1*(2), 85–95. https://doi.org/10.23887/pips.v1i2.2828

Mliless, M., & Larouz, M. (2018). An ecolinguistic analysis of environment texts in Moroccan English language teaching textbooks. *International Journal of Research in Environmental Studies, 5*, 103–116.

Morón-Monge, H., Hamed, S., & Monge, M. del C. M. (2021). How do children perceive the biodiversity of their nearby environment: An analysis of drawings. *Sustainability (Switzerland), 13*(6), 1–19. https://doi.org/10.3390/su13063036

Niswatin, N., & Mahdalena, M. (2016). Nilai Kearifan Lokal "Subak" Sebagai Modal Sosial Transmigran Etnis Bali. *Jurnal Akuntansi Multiparadigma, 6*, 171–188. https://doi.org/10.18202/jamal.2016.08.7015

Parwata, A. A., Oka, G., & Sudantra, I. K. (2018). Subak landscape management as a justice and welfare oriented world culture heritage. *JL Pol'y & Globalization, 80*(1), 11–19. https://heinonline.org/hol-cgi-bin/get_pdf.cgi?handle=hein.journals/jawpglob80&section=5

Pratiwi, Y., Andayani, K., Roekhan, & Prastio, B. (2022). Environmental themes in BIPA textbook: Ecolinguistics perspective. *Proceedings of the International Seminar on Language, Education, and Culture (ISoLEC 2021), 612*(ISoLEC), 323–333. https://doi.org/10.2991/assehr.k.211212.061

Puspawati, D. A., Surata, S. P. K., & (Universitas Mahasaraswati Denpasar). (2018). Ekopedagogi Berbasis Lanskap Budaya Subak: Sumber Belajar Alternatif Dalam Menunjang Ekowisata Dan Pengelolaan Lingkungan. *Prosiding Seminar Nasional Pendidikan Biologi*, 186–190.

Reynolds, Jan. (2013). *Cycle of rice, cycle of life: A story of sustainable farming*. Lee & Low Books.

Risna, R. A., Rustini, H. A., Herry, Buchori, D., & Pribadi, D. O. (2022). Subak, a nature-based solutions evidence from Indonesia. *IOP Conference Series: Earth and Environmental Science, 959*(1). https://doi.org/10.1088/1755-1315/959/1/012030

Ronny, P., & Mahendra, A. (2013). Tri Hita Karana Sebagai Landasan Memperkuat Kepemimpinan Pancasila Putu Ronny Angga Mahendra. *Seminar Nasional INOBALI 2019 Inovasi Baru Dalam Penelitian Sains, Teknologi Dan Humaniora 222*, 3, 222–228.

Roth, D. (2011). The subak in diaspora: Balinese farmers and the subak in South Sulawesi. *Human Ecology, 39*(1), 55–68. https://doi.org/10.1007/s10745-010-9374-7

Ruslin, R., Mashuri, S., Sarib, M., Rasak, A., & Alhabsyi, F. (2022). Semi-structured interview: A methodological reflection on the development of a qualitative research instrument in educational studies Ruslin. *Journal of Research & Method in Education (IOSR-JRME), 12*(1), 22–29. https://doi.org/10.9790/7388-1201052229

Salamanca, A. M., Nugroho, A., Osbeck, M., Bharwani, S., & Dwisasanti, N. (2015). *Managing a living cultural landscape: Lessons and insights from the subaks of Bali, a UNESCO World heritage site*. Stockholm Environment Insitute, 54.

Shepardson, D. P., Wee, B., Priddy, M., & Harbor, J. (2007). Students' mental models of the environment. *Journal of Research in Science Teaching, 44*(2), 327–348. https://doi.org/10.1002/tea.20161

Sriartha, I. P., Jampel, I. N., Widiana, I. W., & Wesnawa, I. G. A. (2017). Local wisdom of subak as a model of character building for social studies learning in schools. In Advances in Social Science, Education and Humanities Research *134*(Icirad), (pp. 114–120). Atlantis Press. https://doi.org/10.2991/icirad-17.2017.22

Sriartha, I. P., Suratman, S., & Giyarsih, S. R. (2015). The effect of regional development on the sustainability of local irrigation system (A case of subak system in Badung regency, Bali Province). *Forum Geografi, 29*(1), 31–40. https://doi.org/10.23917/forgeo.v29i1.789

Sriyanti, N. N. M., & Sriartha, I. P. (2019). Pengembangan Sumber Belajar Ips Berbasis. *Pips, 3*(2).

Subramaniam, K. (2019). An exploratory study of student teachers' conceptions of teaching life science outdoors. *Journal of Biological Education*, 53(4), 399–411. https://doi.org/10.1080/00219266.2018.1472133

Surata, S. P. K. (2013). Pembelajaran Lintas Budaya: Penggunaan Subak sebagai Model "Ecopedagogy." *Kajian Bali, 03*(September), 24–25.

Surata, S. P. K., Jayantini, G. A. S., & Lansing, J. S. (2014). Exploring community capital of the Balinese subak cultural heritage: A content analysis of participatory maps. *International Journal of Technical Research and Applications, 2*(7), 28–34.

Surata, S. P., Arjaya, I. B., Paraniti, A. A. I. P., & Hermawan, I. M. S. (2024). Lokakarya kolaboratif tentang integrasi antara sistem subak, kreativitas visual dan chat dan ChatGPT di SMAN 1 Mengwi Badung Bali. *Mitra Mahajana: Jurnal Pengabdian Masyarakat, 5*(3), 270–279.

Wirahyuni, K., Suandi, I. N., Martha, I. N., Sudiana, I. N., & Permana, I. G. Y. (2021). A thorough analysis on the primary school teachers' belief in modeling Balinese local wisdom of Tri Hita Karana. *Alinteri Journal of Agriculture Sciences, 36*(2), 257–264. https://doi.org/10.47059/alinteri/v36i2/ajas21141

Wuellner, M., Vincent, L., & Felts, B. (2017). Environmental mental models of college students. *International Journal of Environmental and Science Education, 12*(2), 105–115.

Zahoor, M., & Janjua, F. (2020). Green contents in English language textbooks in Pakistan: An ecolinguistic and ecopedagogical appraisal. *British Educational Research Journal, 46*(2), 321–338. https://doi.org/10.1002/berj.3579

6

Representation of ecotourism in Morocco

An ecolinguistic analysis of YouTube videos

Mohamed Mliless, Mohammed Larouz and Mohamed Louza

Introduction

In the evolving landscape of global tourism, ecolinguistic analysis offers a critical lens for understanding how discourse shapes perceptions of destinations, particularly in the Global South. Tourism, as a complex socio-economic and cultural phenomenon, intersects with issues of social justice, environmental sustainability and linguistic representation. Morocco, with its rich touristic heritage spanning historic cities like Marrakech and Fez, and diverse landscapes such as the Atlas Mountains, presents a compelling case study for examining how tourism discourse constructs and portrays complex socio-environmental dynamics. While traditional tourism development strategies have emphasized economic growth, recent scholarship reveals critical nuances in representation and communication strategies. A study by Lamnadi (2019) highlights significant gaps in current public policy and marketing approaches, building upon earlier critiques by Vellas and Mehadji (2005) that challenged conventional tourism management policies.

Technological advancements have fundamentally transformed tourism communication, with social media platforms becoming pivotal spaces for destination representation. Ait Lhou (2017) underscores the profound impact of digital technologies on advertising, particularly in Morocco, while Ait Hattani and Khoumssi (2020) emphasize the crucial role of social networking sites (SNS) in shaping destination narratives. The emergence of environmentalism and increased awareness of cultural sustainability have further complicated these representational strategies, revealing tensions between economic imperatives and ecological consciousness.

This ecolinguistic analysis explores how YouTube videos produced by foreign influencers and the Moroccan tourism office construct and depict representations of natural and cultural landscapes. By examining the linguistic and visual strategies employed in this digital content, the research aims to unpack how different actors mobilize discourse to engage audiences, problematize sustainability and mediate the complex interactions between tourism, culture and environment. The chapter critically interrogates how linguistic choices, visual rhetoric and narrative structures either reinforce or challenge the traditional commodification of cultural and natural spaces, offering insights into the power of language in constructing touristic imaginaries.

Review of the literature

Ecotourism, defined as travelling to relatively unspoiled natural destinations to appreciate nature, learn about wildlife and experience authentic local cultures while preserving the environment of the destination, has been a subject of numerous studies (for a review, see Khanra et al., 2021). Four key themes have emerged from these studies: ecological preservation, the interests of local residents, carbon footprint and tourist behaviour.

The ecolinguistic approach provides a vital framework for understanding the interplay between language, ecology and social dynamics, particularly in the context of tourism and its implications for sustainability. Four key themes have emerged from recent studies in this field: ecological preservation, the interests of local residents, carbon footprint and tourist behaviour. In terms of ecological preservation, ecolinguistics emphasizes the importance of language in shaping perceptions and narratives surrounding ecological preservation. It critiques dominant discourses that promote unsustainable practices while advocating for language that fosters environmental stewardship (Stibbe, 2017). By analysing how ecological issues are framed in public discourse, researchers can identify narratives that either support or undermine conservation efforts. For instance, studies mentioned in a literature review article recently published by Ponton (2024) have shown that language can influence public attitudes towards environmental issues, highlighting the need for positive ecological narratives that promote sustainable practices.

Similarly, the interests of local residents are often marginalized in tourism discourse. For instance, ecolinguistic analyses reveal how language reflects and constructs the socio-cultural realities of these communities. Francis (2024) and Jessen et al. (2023) examined local lexicons and narratives, and uncovered the ecological wisdom embedded within indigenous knowledge systems. This understanding is crucial for ensuring that tourism development aligns with the aspirations and needs of local populations, fostering a more equitable distribution of benefits derived from tourism activities (Aref et al., 2010).

In the same vein, and in relation to the connection of discourse and tourism, carbon footprint has been addressed by ecolinguistics who examined how language shapes awareness and attitudes towards climate change. Khatibi (2021) and Hugé et al. (2013) argue that the framing of carbon emissions in public discourse can significantly impact individual and collective behaviours related to sustainability. Equally important, other ecolinguistic studies (Rudd et al., 2020; Zhang, 2022; Radovanović & Pešić, 2024) have investigated how different linguistic representations of carbon footprints influence public engagement with environmental policies and practices. In terms of tourist behaviour, the ecolinguistic approach has shed light on tourist behaviour by analysing how language influences perceptions of ecotourism. For Laajini et al. (2023), the narratives constructed around ecotourism can either encourage responsible travel practices or perpetuate harmful behaviours. By critically examining the language used in marketing materials and social media, researchers can identify discourses that promote sustainable tourism practices versus those that contribute to ecological degradation (Fill & Mühlhäusler, 2001). This analysis is vital for developing effective communication strategies that encourage environmentally responsible behaviour among tourists.

However, ecotourism faces the important challenge of achieving a balance between environmental and community needs. That is to say, economic growth through tourism should benefit local communities rather than solely external investors, and the number of visitors needs to be managed to minimize their detrimental effects on the environment and wildlife.

The Covid-19 pandemic has significantly impacted the global tourism industry. Before the pandemic, travel and tourism contributed 10 per cent to the global GDP and provided over 320 million jobs (Khanra et al., 2021). Evidently, the pandemic has put 100 million jobs at risk and led to a global decline in tourist arrivals of over 65 per cent in the first half of 2024. The pandemic has also disrupted income streams for many protected areas, leading to financial crises, staff layoffs and reduced law enforcement, which can increase poaching and encroachment.

Types of tourism

Sustainable tourism refers to sustainable practices in and by the tourism industry, aiming to acknowledge all impacts of tourism, both positive and negative, and to minimize the negative impacts and maximize the positive ones. It emphasizes responsible travel practices that benefit local communities, promote cultural understanding and conserve natural habitats (Sofronov, 2017). There are many types of sustainable tourism, often with overlapping definitions, such as ecotourism, green tourism, cultural tourism, responsible tourism and solidarity tourism (Tureac 2008; Sarfaraz et al., 2015). Achieving sustainable tourism is a continuous process that requires constant monitoring of impacts and introducing the necessary preventive and/or corrective measures whenever necessary.

Sustainable and ecotourism

Sustainable tourism, as defined by the World Tourism Organization (WTO) and Chatterjee and Dsilva (2021), is a balanced approach that considers economic, social and environmental impacts, optimizes resources and conserves biodiversity. It respects host communities' socio-cultural authenticity and contributes to inter-cultural understanding, as studied by Idrissi (2023) in Morocco and its alignment with the Sustainable Development Goals.

For instance, green tourism, one form of sustainable tourism, promotes non-degrading travel to natural areas, with natural parks playing a crucial role in environmental protection, biodiversity maintenance and conservation initiatives (Selkani, 2020a). Ecotourism, another form of sustainable tourism, focuses on responsible travel to natural areas, aiming to conserve the environment, improve local well-being, educate travellers and fund conservation. It involves local residents in community projects, limits tourist numbers for sustainability, and uses advertising and social media for promotion (Chatterjee & Dsilva, 2021). Selkani (2020b) draws on the work of Dr Hassan Aboutayeb, a pioneer of nature tourism in Morocco, and argues in favour of ecotourism as a social solution that balances preservation with economic and social development.

Role of media and social network sites

Tourism research, though seemingly new (Reisinger, 2015), has roots in Victorian England and the Grand Tour, associating leisure travel with personal growth and exploration. More recently, the new media revolution has significantly influenced the tourism industry by creating opportunities for user engagement and decision-making behaviours through online communities (Ait Hattani & Khoumssi, 2020). It facilitates social interaction and collaboration among travellers, allowing service providers to improve their productivity and revenues through consumer feedback (Ramos et al., 2019).

The media, including television, print, online platforms and social media, plays a pivotal role in promoting ecotourism by raising awareness, disseminating information and shaping public perceptions (Blackshaw, 2006). It communicates the principles and benefits of ecotourism, raises conservation awareness and showcases natural landscapes. The media also highlights success stories and innovative practices in sustainable tourism, inspiring audiences to make responsible travel choices.

A recent study by Anfoud and Karim (2020) highlighted the role of digital technology in improving the attractiveness of territories. They proposed a digital marketing strategy based on attractiveness, conversion and loyalty, which help improve a site's web visibility. The study emphasized the potential of social networking sites (SNS) in enhancing the value of ecotourism by creating high-quality content on an official website and sharing it on an SNS. This approach can strengthen a region's brand image, develop its reputation and increase its appeal. The chapter also underscored the importance of examining how various stakeholders, such as officials or influencers, promote tourism in Morocco.

Moroccan visions: 2010 and 2020

Morocco previously put forward two deadlines for tourism visions focusing on sustainability and ecotourism, aiming to enhance the competitiveness of the Moroccan tourism industry by addressing structural weaknesses and aligning with new trends and opportunities (Roudies, 2013). The '*2010–2020 Vision*' strategy, established in 2010 by the Moroccan government and the General Confederation of Moroccan Enterprises (CGEM), first aimed to attract 10 million tourists and for tourism to represent 20 per cent of the national GDP by

2010 (Ait Hattani & Khoumssi, 2020). This strategy led to an increase in tourist arrivals from 4.3 million in 1999 to 9.2 million in 2010, ranking Moroccan tourism 25th worldwide (Bouzahzah & El Menyari, 2013).

The second goal was to make Morocco one of the world's top twenty tourist destinations and a model for sustainable tourism in the Mediterranean area by 2020. The strategy aimed to double tourist arrivals, increase tourism receipts to about 140 billon Moroccan Dirhams (MAD) in 2020 and triple domestic trips. In this regard, WTO forecasted that by 2020, the number of tourist arrivals in Africa more generally would reach 77 million (Roudies, 2013).

Methodology

The representation of ecotourism in media and social networks in Morocco is a complex issue that requires a sophisticated methodology to analyse both discourse and visuals. The portrayal of ecotourism on SNS can significantly impact public perception, and a comprehensive analysis of visuals and discourse can provide insights into its promotion and perception (Kress & Van Leeuwen, 2006). An online sentiment analysis tool, namely Daniel Soper's 'Free Sentiment Analyzer', was used to evaluate the sentiment or tone of video comments, assigning a sentiment score ranging from positive (0 to 100) to negative (0 to −100).

The chapter assumes that analysing the characteristics and benefits of ecotourism, as defined by organizations like WTO, and the marketing strategies used in a promotional video can enhance understanding of its effectiveness. Characteristics include nature-based tourism with educational features, organized by specialized tour operators for small groups, with service providers being small, locally owned businesses. The video should minimize negative environmental impacts and support the maintenance of natural areas.

The analysis will look for images that effectively communicate ecotourism's characteristics and benefits, such as nature-based tourism activities, small groups led by local guides and efforts to minimize negative impacts. In terms of comments, the chapter will identify linguistic elements that reflect these characteristics and values, such as specific keywords or phrases like 'nature-based', 'sustainable development' or 'local communities', and the tone and sentiment of the comments.

Theoretical framework: Ecolinguistics

The importance of discourse in addressing ecological disasters is undeniable, as texts, handbooks, lifestyle magazines and economics textbooks can impact human behaviour, even if they are not explicitly defending an 'ecological' discourse (Stibbe, 2015). The scope of ecology within linguistics has broadened since the early days of the field. Halliday (1990), a pioneer in ecolinguistics, analysed the language of ecological texts, arguing that certain grammatical aspects conspire to construe reality in a way that is no longer good for our health as a species. Chalwa (1991) focused on the grammar of languages, claiming that the language habits of fragmenting the mass, quantifying intangibles and imaginary nouns, and perceiving time in terms of past, present and future are factors in our inability to perceive the natural environment holistically. Both Halliday and Chalwa preferred analysing discourses within an ecological framework rather than restricting themselves to examining discourses which happen to be about ecology.

Within these dynamics, ecolinguistics, a field of study for over fifty years, has emerged to explore the interactions between language and its environment (Mühlhäusler & Fill, 2001). It started as a united stream of linguistics in the early 1990s, connecting language and ecology (Derni, 2008). Mühlhäusler and Fill (2001) outlined its three main principles: recognizing and defending diversity, recognizing mutual interaction, and perceiving wholeness and unity rather than fragmentation, summarized as 'diversity + interaction = wholeness and unity'.

Ecolinguistic perspectives and approaches

Stibbe (2015) emphasized that ecolinguistics explores linguistic phenomena from an ecological perspective, revealing and questioning texts, contributing to the search for new ones, and examining ideologies. It provides tools for analysing everyday texts, uncovering hidden stories and evaluating them from an ecological perspective to determine whether they encourage the protection or destruction of ecosystems.

Discourse analysis has been a key tool in environmental policy since the mid-1990s, with a focus on ecological modernization (Leipold et al., 2019). The identification of four main discourses – 'problem-solving', 'sustainability', 'survivalism', and 'green radicalism' – has led to a range of studies on topics from acid rain to climate change. These included the policies of agriculture, biodiversity, climate governance, environmental justice, forest policy discourses, nature conservation, renewable energy, sustainable development and climate-conflict-displacement issues (Behnassi et al., 2022).

Ecolinguistic research and applications

Ecolinguistic scholars worldwide are committed to building theoretical and methodological resources to reveal the role of discourse in describing, concealing or solving ecological problems (Zhang, 2022). They have analysed various texts, including textbooks, fables, religious verses, news media coverage of climate change, dominant discourses, advertisements and the discourse of economy and business (Zhang, 2022). Furthermore, scholars have discussed how ecology appears in different languages worldwide.

Ecolinguistics in Morocco is an emerging trend with scholars investigating environmental themes using frameworks of ecolinguistics, discourse analysis, semiotics, and media perspectives among others. Several studies have examined environmental discourse in various contexts, such as English-language teaching textbooks, eco-documentary narratives, fear expressions in a Moroccan eco-documentary, language and visuals in a Moroccan public service announcement for forest protection, and coverage of environmental issues in print media during the Covid-19 pandemic (Mliless et al., 2021).

From an ecolinguistic perspective, research has shown how linguistic strategies can reveal underlying ideologies when evaluating a tourism website. In terms of tourism narratives, a study on Kalimantan tourism texts found that appraisal patterns, expressed through attitude and graduation systems, created positive feelings in readers due to the quality and quantity of the Kalimantan environment (Istianah & Suhandano, 2022). However, the study also noted an ambivalent discourse: while the website promoted natural richness to raise awareness of Kalimantan's role as the world's lungs, it also encouraged the objectification and commodification of nature in the tourism sector (Istianah & Suhandano, 2022).

Visual approaches to ecotourism: Sociosemiotics

Research on tourism images has explored various media such as commercials, postcards and photos. Studies have analysed the effects of promotional videos on destination image, components of attractive tourism destination images and differences in content structure between agency-generated and user-generated videos. Investigations have also been conducted into the role of tourism in developing countries, the impact of agricultural landscape externalities on rural poverty and the use of consumer research results in creating advertising messages for artisanal tours (Karimova, 2015).

The creation of meaning, essential to human communication and living, is embodied in signs. Semiotics, the science of signs, emerged with contributions from individuals like Ferdinand de Saussure and Charles Sanders Pierce. Social semiotics, often associated with Halliday (2007), emphasizes the importance of understanding language to understand human beings as social creatures. It recognizes the symbiotic relationship between the sign, sign maker/user and social context. The social semiotic theory developed by Gunther Kress and Theo Van Leeuwen (2006) extends Halliday's concepts about language to the analysis of visual images.

Data description

This chapter analyses seven YouTube videos in order to gain a deeper understanding of the representation of ecotourism in Morocco. The analysis focuses on both the visual content of the videos and the language used in the accompanying comments. The visual content provides important insights into how ecotourism is being represented and promoted, while the language of the comments reveals viewers' perceptions and responses to the videos. By examining both aspects, this work seeks to provide a deeper understanding of the representation of ecotourism in social networks. Data was taken from four personal travel videos on YouTube: *Come with us*, *J'irai dormir*, *On va finir*, and *Tu ferais*, as indicated in Table 6.1. Similarly, we opted for three videos by the Moroccan Office of Tourism (MOT) on the promotion of ecotourism from an official perspective: *Découvrez Bab Louta*, *Découvrez Taza*, and *Découvrez Tafoughalt*. The analysis of these videos unveils how viewers of ecotourism perceive and react to the ecological destinations as promoted by both foreign influencers and the MOT.

In quantitative terms, the numbers in Table 6.1 represent statistics for seven videos in the two categories. The first column represents the two types: personal travel videos (henceforth 'travel videos' vs. government-sponsored videos from

Table 6.1 Videolinks to Tourist Videos on YouTube from Personal Travel vs. Government Sources

Type		Title	Time	Views	Comments	Likes	Dislikes	Web links
Personal travel videos	**1**	*Come with us to The Kingdom of Morocco*	14 m 10 s	471,909	468	4.4k	**0**	https://www.youtube.com/watch?v=Ko1MQnJ40wE&feature=youtu.be
	2	*J'irai dormir chez vous au Maroc*	52 m 23 s	709,776	1,419	5.6k	**0**	https://www.youtube.com/watch?v=YJ_4JKAnbUA
	3	*On va finir par s'installer au Maroc à ce rythme*	26 m 18 s	1,550,159	4,481	56k	0	https://www.youtube.com/watch?v=43y-40ytED8
	4	*Tu ferais confiance à un inconnu toi? – Rencontres au Maroc*	25 m 44 s	1,716,333	8,806	97k	0	**https://www.youtube.com/watch?v=QNTTY_goU7w&t=572s**
MOT videos	**5**	*Découvrez Bab Louta, en partenariat avec l'ONMT*	4 m 05 s	2,932	0	62	0	https://www.youtube.com/watch?v=kKmbuFXLNdM
	6	*Découvrez Taza, en partenariat avec l'ONMT*	4 m 05 s	548	0	9	0	https://www.youtube.com/watch?v=KbX0BW7CfCQ
	7	*Découvrez Tafoughalt, en partenariat avec l'ONMT*	4 m 31 s	735	0	6	0	https://www.youtube.com/watch?v=dn8IzSoRatc

the MOT. The second column represents the duration of the video in minutes and seconds; the third column represents the number of views the video has received; the fourth column represents the number of comments on the video; the fifth column represents the number of likes for the video; and the sixth column represents the number of dislikes for the video. The video with the most views is *Tu ferais* with 1,716,333 views, followed by *On va finir* with 1,550,159 views, *J'irai dormir* with 709,776 views and *Come with us* with 471,909 views. The video with the most comments is *Tu ferais* with 8,806 comments, followed by *On va finir* with 4,481 comments, *J'irai dormir* with 1,419 comments and *Come with us* with 468 comments. None of the videos have any dislikes.

As for the second category of videos, produced by the MOT, the numbers represent statistics for three videos, and each video has duration of around four minutes. *Bab Louta* has the most views with 2,932, followed by *Tafoughalt* with 735 views and *Taza* with 548 views. However, none of the videos have any comments. *Bab Louta* also has the most likes with 62, followed by *Taza* with 9 likes and *Tafoughalt* with 6 likes. Similarly, none of the videos have any dislikes.

Results

An analysis of the titles of each video suggests that the American and French influencers focus on personal experiences and interactions with locals, while the MOT videos focus on promoting specific locations in Morocco. When the focus is on a particular region, the videos aim to provide information about the natural beauty, history and culture of that specific area. However, if the aim of the videos is to focus on a region rather than on the community, the viewer may miss out on information about the local people and their customs, traditions and way of life. The videos can help to create a more personal connection between the viewer and the region, which can be a powerful motivator for people to visit. Social media platforms, in particular, extend the reach of ecotourism messages by allowing people to share their personal experiences and forming an online community dedicated to sustainable tourism.

Video 1: '*Come with us to the Kingdom of Morocco*'

Sociosemiotic analysis

This video highlights the importance of preserving the unspoiled beauty of the landscape in promoting sustainable tourism. Visitors appreciate the opportunity to immerse themselves in nature and share their experiences through photographs and videos. Captivating images and videos of eco-friendly practices, customer testimonials and sharing eco-friendly initiatives on social media can build trust and appeal to environmentally conscious travellers. In video 1, ecotourism is displayed as responsible travel to natural areas that conserves the environment, sustains the well-being of local people, and involves interpretation and education. It also raises environmental awareness, supports conservation efforts and fosters respect for different cultures as well as natural habitats.

The research indicates that video content can effectively promote ecotourism when it incorporates key elements such as nature-based tourism and support for the conservation of natural areas. However, it is found that these characteristics were weakly represented in the videos, which could have potential implications for all sectors involved uplifting ecotourism. A weak or negative representation of ecotourism in videos could lead to a lack of understanding or interest among viewers, decreased demand for ecotourism opportunities, and negative impacts on local communities and businesses. Therefore, it is important for videos promoting ecotourism to effectively communicate its benefits to encourage engagement and support for this form of sustainable tourism. This can help support the maintenance of natural areas, minimize negative impacts on the environment, provide economic benefits and empower local communities.

Connection with nature and local communities

Video 1 (YouTube Channel Link: Table 6.1) is significant in promoting ecotourism by helping people recognize and appreciate the unique ecosystems of different destinations. It highlights this through stunning photographs and videos; viewers can gain a deeper understanding of the natural beauty and diversity of these environments, inspiring them to visit and support ecotourism initiatives. The use of images also plays a crucial role in helping visitors connect with nature, inciting people to experience the beauty and wonder of the natural world, even before they visit a destination. These images can evoke a sense of awe and appreciation for the environment, inspiring visitors to seek out opportunities to immerse themselves in nature and engage in ecotourism activities. The power of guest photos and videos is also emphasized as an effective tool for marketing the sustainable tourism aspects of businesses, as they showcase the sustainable ethos of tourism businesses and create a sense of community, encouraging appreciation of craftsmanship and highlighting authenticity.

Happiness, joy, entertainment and emotional fulfilment

Videos on ecotourism in Morocco can play an important role in showcasing the happiness, joy, entertainment and emotional fulfilment that visitors experience when they come to the country. Through this video, viewers can see first-hand the positive impact that ecotourism can have on both the local community and the environment, as well as the personal satisfaction and enrichment that visitors gain from engaging in responsible travel practices. By capturing the smiles, laughter and excitement of visitors as they explore the natural beauty and cultural heritage of Morocco, videos can inspire others to visit the country and experience the same sense of happiness and fulfilment.

Video 1: Sentiment analysis of comments

This chapter adopts sentiment analysis as a technical instrument to evaluate comments on YouTube videos about ecotourism. Sentiment analysis, a technique that determines the emotional tone behind words, can be used to classify comments as positive, negative or neutral. For this aim, we adopted Daniel Soper's Sentiment Analyser free tool, to gauge viewer attitudes. The tool calculates a sentiment score representing the overall sentiment, tone or emotional feeling conveyed by the text. The scores range from −100 to +100, with −100 indicating a very negative or serious tone, and +100 indicating a very positive or enthusiastic tone. Soper's Sentiment Analyser free tool provides insights into viewer responses to the video content to understand the emotions, attitudes, evaluations and opinions expressed by users.

The first video by American tourists, which praises ecotourism localities in Morocco, has received a lot of positive feedback from viewers. Many have expressed their love for Morocco and its beauty, hospitality and affordability. Some viewers have even been motivated to visit Morocco after watching the video. The breathtaking landscape scenery and beautiful waterfalls have been particularly appreciated. Overall, the video seems to have left a lasting impression on its viewers. The comments on the video by American tourists were submitted to Daniel Soper's Sentiment Analyser free tool. The tool analysed the overall sentiment or tone of the comments and gave them a sentiment score of 80.4. This means that the overall sentiment of the comments is very positive and enthusiastic. This indicates that the viewers had a very favourable impression of Morocco and its ecotourism localities.

Video 2: *'J'irai dormir chez vous'/'I will sleep at your place'*

Sociosemiotic analysis

Antoine de Maximy, a French video producer born on 21 May 1959 in Lyon, is renowned for his show *J'irai dormer chez vous* (YouTube Channel Link: Table 6.1). He is a multifaceted personality – a backpacker, television host, wildlife photographer, filmmaker, journalist and war correspondent. During a visit to Morocco, he explored and documented a deserted locality, once a thriving lead mine, travelling on foot and by motorcycle.

Ecotourism and hospitality: A glimpse into Morocco's forgotten lead mine

De Maximy's video sheds light on the history and culture of a remote area near a deserted lead mine in Morocco. The story revolves around his encounter with Mbarek, a guard at the mine, and his family. The video not only highlights the diverse ecosystem and wildlife of the area but also emphasizes the hospitality and generosity of the local community. Furthermore, it advocates for ecological tourism to promote the area and improve the socio-economic conditions of the residents, such as Mbarek's family and their activities in the village. The narrative showcases the potential for ecotourism to benefit the local community while offering visitors a unique opportunity to learn about the area's history and culture as seen in this exchange:

ANTOINE Do you want me to pay a little for the accommodation if I stay here tonight? Yes, you'll give me food . . . I'll give you a little money.

MBAREK No money, no nothing. You sleep, you eat, and you go; safe travel and safe travel.

Video 2: Sentiment analysis of comments

As regards video 2, *J'irai dormer chez vous*, the sum of comments yielded a highly positive response towards Morocco and its people. Particularly, the interpretation of the text states that it has a sentiment score of 83.1, indicating that the overall tone of the text is very positive or enthusiastic. This positive sentiment score is reflected in the comments, which praise the kindness, hospitality and generosity of Moroccans, and describe Morocco as a wonderful and welcoming country that is worth visiting.

Video 3: *'At this rate, we'll end up settling in Morocco'*

Sociosemiotic analysis

Our analysis of the French influencers, namely Olivia Kohler (Figure 6.1), on the promotion of ecotourism in Morocco allows us to understand that values such as smiling, sharing, hospitality, knowledge, generosity, help and trust were not only expressed through images and scenes but were also articulated in the narratives. It is a method that we call the articulation of narratives and images on screen (ANIS). This approach allows for a deeper understanding of the values and messages being conveyed through the combination of visual and verbal storytelling.

The experience of the French influencers expressed in video 3 (YouTube Channel Link: Table 6.1) strongly relates to ecotourism, which entails responsible travel to natural areas, conserving the environment and sustaining the well-being of local communities. By sharing their experiences through videos, the French girls hope to provide interpretation and education, key components of ecotourism. On their way to meet a very special man named Mohamed, they coined the expression 'The Smile of Skoura', and it's easy to see why. He is a true incarnation of kindness, as are most of the people they meet on their journey: 'they have their hearts on their hands and possess amazing know-how'. We suggest that this is a beautiful aspect of the journey that attempts to promote the splendour of the region of Skoura.

Figure 6.1 Olivia Kohler: French journalist, photographer and content creator.

Figure 6.2 Signs of generosity.

The power of the smile and generosity

Smiling is a universal gesture that conveys friendliness, happiness and warmth. When receiving tourists, smiling can help to create a welcoming and hospitable atmosphere. By smiling and being friendly, hosts can help to ensure that tourists have a positive travel experience and feel more connected to the local community. This can contribute to the goals of ecotourism by promoting greater understanding and appreciation for nature, local society and culture. In this situation, the girls mentioned that 'Mohamed's smile is contagious. He was delighted to be with us and shared many stories'. Generosity is a valuable trait that can enhance the experience of tourists (Figure 6.2). In the context of ecotourism, generosity can play an important role. By being generous with their time, resources and knowledge, hosts can help tourists feel more connected to the local community and culture. This can contribute to the goals of ecotourism by promoting greater understanding and appreciation for nature and local society. Thus, the narratives in the video refer to Mohamed's immeasurable generosity: 'His generosity and happiness were evident.'

Pillars of ecotourism: Help, incredibility, sharing and trust

This video explicitly calls for people to visit the place in order to help the community cope with the challenges that Covid-19 imposed on craftsmen.

During the pandemic, many people faced hardships and struggles. One such person is Mohamed, a talented pottery maker known as 'The Smile of Skoura'. Due to the pandemic, his passion for pottery was put on hold for many weeks. Without visitors, he was unable to work as usual and could not make enough sales to support his family. Mohamed had to work as a mason for a while until he could return to his occupation. Despite these challenges, he remains motivated by the potential succession of his daughter. He loves passing on the knowledge he learned from his father and grandfather before him, keeping the family tradition alive, 'Let us come together to support him [Mohamed] and others like him who have been impacted by the pandemic.'

In the context of ecotourism, the incredible nature of the hosting community includes the unique and exceptional qualities or experiences that the community has to offer to foreign visitors. These include natural beauty, cultural heritage, local traditions and other attractions that make the community stand out as a tourist destination. By showcasing their incredibility, the hosting community can attract tourists and provide memorable and enriching travel experiences. Again, the video cites Mohamed as truly amazing, 'clearly one of the most beautiful encounters we have had in Morocco'.

The beautiful encounters and sharing of daily details continue to charm the visitors. For the hosts, it can provide an opportunity to showcase their culture, traditions and way of life. This can help to promote greater understanding and appreciation for their community, and can contribute to the preservation of their cultural heritage. For tourists, experiencing the daily life of the local community can provide a unique and authentic travel experience. It can help them to gain a deeper understanding of the local culture and to form more meaningful connections with its people: 'Hakim, Brahim, Sofiane, Mohamed, and Hassan are all wonderful people who have shared a part of their daily lives with us and have become part of our loving family.'

Trust is important for a hosting community that receives tourists for several reasons. First, trust helps establish a positive relationship between the hosts and the tourists. When tourists trust their hosts, they are more likely to feel more comfortable in their new environment (Figure 6.2). This can enhance their travel experience and encourage them to return in the future. Second, trust might also help promote responsible and sustainable tourism. When tourists trust their hosts, they are more likely to follow local customs and guidelines, respect the environment and support local businesses. All these ensure that tourism has a positive impact on the hosting community and contributes to its long-term sustainability. Finally, trust can also establish the hosting community as a tourist destination. When tourists have positive experiences and trust their hosts, they are more likely to recommend the destination to others: 'We have forged real bonds of friendship and discovered many beautiful places and stories. We trust them and share beautiful moments.'

Knowledge sharing and hospitality: Key aspects of ecotourism

Sharing knowledge is important for a hosting community that receives tourists for several reasons. For instance, by sharing their knowledge and expertise with tourists, hosts can provide them with a deeper understanding of the local culture, history and environment. Another reason is that sharing knowledge can also help to promote responsible and sustainable practices. That is to say, tourists do not only learn about local customs, traditions and environmental concerns, but they are also encouraged to adopt responsible and friendly attitudes. Finally, when tourists have positive experiences and learn from their hosts, they are more likely to recommend the place to others. For instance, in this video they praised local knowledge about granaries or 'old bank systems' built in inaccessible areas that are 'always well protected by a community or natural border and are either like this or in caves. It's not an easy place to go, but it's fascinating to learn about.'

Finally, the video's narrative voice ends with hospitality, which in Morocco is expressed in many ways. Moroccans are known for their friendliness and warm hospitality. They provide their guests with everything they need and show them that they are welcome through the food they share with them. It is considered shameful for guests to finish their dish because there should always be too much food. It is also considered shameful to present a dish without meat to guests. Offering tea is a sign of good hospitality and friendship. Moroccans also have a unique way of serving the sweet drink, raising the teapot higher and higher from the glasses while pouring. Moroccan families may invite travellers into their homes after a conversation, and the invitation is usually sincere and the host will be glad to receive you. Hosting in Morocco remains a complicated long process where the host goes to massive expenses and does their best to provide a complete experience: a great meal, an interesting conversation, and a unique area of rest and relaxation. Guests may even be treated like family: 'Our host even said, "This is like your home here!" and his wife said we were like her daughter.'

Video 3: Sentiment analysis of comments

The same process was undertaken to interpret the comments of the third video (see link in Table 6.1). According to the tool, the collection of comments yielded a highly positive response towards Morocco and its people. In particular, the interpretation states that it has a sentiment score of 86.7, indicating that the overall sentiment of the text is very positive or enthusiastic. Accordingly, results are positive.

Video 4: *'Would you trust a stranger?'*

Sociosemiotic analysis

Video 4 highlights an expedition by two French women who aimed to make meaningful encounters, discover new cultures and document their experiences (YouTube Channel Link: Table 6.1). The chapter underscores that preconceived notions about Morocco, often derived from social media representations of Marrakech and the Merzouga dunes, can be broadened by exploring the country more thoroughly, revealing its authenticity.

This travel video documents an encounter with Mohamed, a local shepherd. The French explorers' comment 'When we come with him, we take the time to talk and the feeling passes right away.' They express gratitude for the opportunity to 'live these moments and discover the daily routines of the people'. Open to the unknown, they initially maintained caution as 'two girls alone in the middle of nowhere', but through conversation, they discovered that 'in fact they always had a good intention'. After initial suspicion, they engaged in exchanges with the stranger, leading to experiences that 'you wouldn't necessarily experience if you sleep in a hotel' (Figure 6.3).

The experience showcases the hospitality of the Moroccan people and their openness to sharing their culture with foreigners. The girls were invited by Ismail, Mohamed's friend, to have tea. Despite Ismail's limited French proficiency and the girls' limited knowledge of Berber, they found ways to communicate beyond words, such as through gestures and drawings on the soil and rocks. This highlights that shared experiences and nonverbal cues can sometimes convey as much as words, demonstrating the universal human ability to connect and understand each other despite language barriers.

The French tourists' experience in the Atlas Mountains highlighted the remarkable hospitality of the local people. Despite their initial doubt towards offers of food, the tourists discovered that Mohamed was not selling bread and apples but freely giving them. His generosity reflects the deeply held Moroccan values, centred on sharing and exchange (Figure 6.4). The tourists learned that

Figure 6.3 Total trust.

Figure 6.4 First encounter and trust.

refusing these gifts or offering money in return would be considered impolite. These encounters underscored that giving and sharing are heartfelt actions, integral to Moroccan society (Figure 6.5).

Reflecting on their journey, the French girls recount their unexpected encounter with two strangers in the mountains. Initially, they doubted the possibility of connecting with people they had never met before. They experienced a whirlwind of emotions and expressed gratitude for the opportunity to share these moments with people who were once strangers but had now left an indelible mark on their lives. The day, intense and rich in emotions and sharing, will always be remembered. Reflecting on the experience as a valuable life lesson, one of the girls expresses her gratitude by saying, 'Thank you, life.'

'Tu ferais: You would do' depicts the experience of French girls and reminds us to approach new experiences with an open mind and avoid preconceived notions. While meeting strangers in a different community can be daunting, it's not always threatening. Openness to new experiences and trust in the kindness

Figure 6.5 The value of sharing.

of strangers led to an unforgettable and transformative experience for the girls. The video exemplifies the celebration of different cultures and the embracing of new experiences. It highlights the unique opportunity to explore natural localities and experience the hospitality and kindness of Moroccans, especially in remote areas often underrepresented in the media.

Video 4: Sentiment analysis of comments

Using Daniel Soper's 'Free Sentiment Analyzer', the analysis of the comments posted by viewers on the video shows that they expressed highly positive attitudes towards Morocco and its people. In particular, the interpretation of the text states that it has a sentiment score of 91.4, indicating that the overall emotional tone of the text is very positive. According to the tool, the collection of comments yielded a highly positive response towards Morocco and its people. In this context, a viewer commented on the credibility of the video and the ability of the two girls to effectively showcase the beauty of nature and the hospitality of its inhabitants: 'Well, you showed us that Morocco is not only Marrakech and Tangier, thank you.'

Results of MOT videos

The Moroccan Office of Tourism (MOT) and the television network 2M have collaborated on a series of videos to promote ecotourism in remote areas. The videos feature the regions of Bab Louta, Taza and Tafoughalt, presented by a female host. The first video showcases Bab Louta, a destination known for its lake, cork oak forest, and opportunities for activities such as kayaking and fishing. The second video highlights the cultural heritage and natural beauty of Taza, including its medina, great mosque and Tazekka National Park. The final video focuses on Tafoughalt, a small town in the mountains with waterfalls, hemp plantations and growing tourist appeal. These videos aim to promote responsible and sustainable tourism in line with Morocco's commitment to ecological responsibility and preservation of its diverse landscapes and wildlife.

Themes promoted by MOT videos

The videos analysed here are based on the depiction of transportation, natural resources, touristic infrastructures, food and culture of the regions. Each video commences with a map that marks a destination for visitors, accompanied by information on the available public transportation to these locations. The videos underscore the natural landscapes of the regions, such as rivers and lakes, and highlight the diversity of local food dishes and accommodation for tourists planning overnight stays. The overarching objective of these videos is to stimulate interest and encourage exploration of these unique and picturesque areas of Morocco.

Efficient transportation systems are important for tourism, as they allow tourists to easily explore a destination, and different transportation options cater to different types of tourists. Videos produced by MOT promote tourism in Morocco by showcasing the country's attractions and encouraging sustainable tourism practices.

The MOT videos promote varied food and culture by a focus on unique cuisine and local dishes. By highlighting the country's culture, the MOT can attract visitors interested in experiencing Morocco's traditions and customs. The videos also encourage tourists to explore the diverse attractions and experiences of each region. Morocco's transport system is expanding, with efficient public transportation and a budget-friendly train network linking major cities. Upgrades to transport and logistics infrastructure aim to boost the sector and the economy, making it easier for local producers and transport services companies to operate.

The MOT's videos promote the natural beauty, transportation options, food and culture of Morocco but neglect to mention the importance of connecting with local communities, supporting their socio-economic development and ecotourism. Ecotourism is a key reason why many visitors choose to travel to these destinations, as it allows them to experience the natural beauty of the region in a sustainable and responsible manner.

The MOT has blocked comments on its ecotourism promotion videos, possibly due to the video creator's decision or a violation of community guidelines. This has prevented the analysis of viewer feedback on the ecotourism promotion videos. Ecotourism, which involves travel to natural areas providing environmental, cultural and economic benefits, is a rapidly growing market in the tourism industry, accounting for 11.4 per cent of global consumer spending. Therefore, it is of the utmost importance that the MOT promote the human dimension of travel in a more appealing way.

Conclusion

This research examines the impact of social media on ecotourism, emphasizing the potential of personal travel videos to support sustainable tourism practices. These videos play a crucial role in effectively communicating the characteristics and benefits of ecotourism, encouraging engagement and support for this form of sustainable tourism. It was found in this chapter that while the Moroccan Ministry of Tourism's videos promote the natural beauty, transportation, food and culture of the regions, they often overlook the importance of connecting with local communities and supporting their socio-economic development. Additionally, the research stresses the significance of promoting responsible travel practices and ecotourism to ensure the protection of natural resources for future generations. Ultimately, ecotourism, through responsible travel practices, aims to create a positive and mutually beneficial relationship between tourism, nature and local communities, fostering sustainable development and safeguarding the well-being of the planet for future generations to enjoy. In the vast landscape of the internet, where countless marketing visuals compete for attention, the choice of the most accurate and impactful visual is essential for the success of any campaign. Based on that, some recommendations for promoting ecotourism through social media could include running targeted advertising campaigns and gathering data to improve messaging; collaborating with bloggers and travel influencers to promote ecotourism; ensuring that video content promoting ecotourism includes key elements such as nature-based tourism with educational features, organized by specialized tour operators for small groups with service providers being small, locally owned businesses;

and minimizing negative impacts on the environment and supporting the maintenance of natural areas.

References

Ait Hattani, H., & Khoumssi, K. (2020). The role of new media in travel and tourism industry in Morocco: Customers' attitudes. *Journal of Media Research, 13*, 2(37), 111–128.

Ait Lhou, D. (2017). La Publicité Touristique, une Motivation Symbolique du Marketing Territorial a Marrakech a L'ère de la Consommation [Tourist advertising, a symbolic motivation of territorial marketing in Marrakech in the age of consumption]. *Revue Marocaine de Recherche en Management et Marketing*, 17, 398–413.

Anfoud, M. & Karim, K. (2020). The attractiveness of territories in the digital era: Case of the city of Marrakech. *Moroccan Journal of Research in Management and Marketing, 12*(2), 60–81.

Aref, F., Gill, S. S., & Aref, F. (2010). Tourism development in local communities: As a community development approach. *Journal of American Science*, 6(2), 155–161.

Behnassi, M., Gupta, H., Kruidbos, F., & Parlow, A. (Eds.). (2022). *The climate-conflict-displacement nexus from a human security perspective.* Springer.

Blackshaw, P. (2006). *The consumer-generated surveillance culture.* Retrieved from www.clickz.com/showPage.html?page=3576076

Bouzahzah, M., & El Menyari, Y. (2013). International tourism and economic growth: The case of Morocco and Tunisia. *The Journal of North African Studies, 18*(4), 592–607.

Chalwa, S. (1991). Linguistic and philosophical roots of our environmental crisis. *Environmental Ethics, 13*(3), 253–73.

Chatterjee, J., & Dsilva, N. R. (2021). A study on the role of social media in promoting sustainable tourism in the states of Assam and Odisha. *Tourism Critiques: Practice and Theory, 2*(1), 74–90.

Derni, A. (2008). The ecolinguistic paradigm: An integrationist trend in language study. *The International Journal of Language Society and Culture, 24*, 21–30.

Fill, A., & Mühlhäusler, P. (2001). *Ecolinguistics: Language, ecology and the environment.* Continuum.

Francis, O. (2024, March 12). The role of indigenous knowledge in environmental conservation. *Medium.* Retreived from: https://olamidefrancis.medium.com/the-role-of-indigenous-knowledge-in-environmental-conservation-f16db06e5a83

Halliday, M. A. K. (1990). New ways of meaning: The challenge to applied linguistics. *Journal of Applied Linguistics*, 6, 7–36.

Halliday, M. A. K. (2007). Language as social semiotic: Towards a general sociolinguistic theory. In J. J. Webster (Ed.), *Collected works of M.A.K. Halliday: language and society*. London: Continuum.

Hugé, J., Waas, T., Dahdouh-Guebas, F., Koedam, N., & Block, T. (2013). A discourse-analytical perspective on sustainability assessment: Interpreting sustainable development in practice. *Sustainability Science, 8*, 187–198. https://doi.org/10.1007/s11625-012-0184-2

Idrissi, F. C. (2023). Sustainable tourism in Morocco: Several actions undertaken with great challenges to fit it into the goals of sustainable development. *International Journal on Optimization and Applications IJOA, 03*(1), 25–31.

Istianah, A., & Suhandano, S. (2022). Appraisal patterns used on the Kalimantan tourism website: An ecolinguistics perspective. *Cogent Arts & Humanities*, 9(1). 1–16. https://doi.org/10.1080/23311983.2022.2146928

Jessen, T. D., Ban, N. C., Claxton, N. XEMŦOLTW, & Darimont, C. T. (2023). The role of indigenous knowledge in environmental conservation. *Ecological Solutions and Evidence, 4*(1), e2435. https://doi.org/10.1002/fee.2435

Karimova, G. Z. (2015). How to transform research results into the advertising message: The example of experiential artisanal tours in Morocco. *Global Journal of Management and Business, 2*(1), 45–53.

Khanra, S., Dhir, A., Kaur, P., & Mäntymäki, M. (2021). Bibliometric analysis and literature review of ecotourism: Toward sustainable development. *Tourism Management Perspectives, 37*, 100777. https://doi.org/10.1016/j.tmp.2020.100777

Khatibi, F. S., Dedekorkut-Howes, A., Howes, M., & Torabi, E. (2021). Can public awareness, knowledge and engagement improve climate change adaptation policies?. *Discov Sustain, 18*(2). https://doi.org/10.1007/s43621-021-00024-z

Kress, G., & Van Leeuwen, T. (2006). *Reading images: The grammar of visual design.* Routledge.

Laajini, T., Lemgadar, A., & Jekki, H. (2023). L'apport de l'écotourisme au développement des territoires: Une analyse par le récit de voyage [The contribution of ecotourism to territorial development: An analysis through travel narratives]. *RIMEC*, 10, Retrieved from: http://revue-rimec.org/lapport-de-lecotourisme-au-developpement-des-territoires-une-analyse-par-le-recit-de-voyage/

Lamnadi, Y. (2019). The evaluation of tourism marketing model within Moroccan public policy management. *Athens Journal of Tourism, 6*(1), 15–38.

Leipold, S., Feindt, P. H., Winkel, G., & Keller, R. (2019). Discourse analysis of environmental policy revisited: Traditions, trends, perspectives. *Journal of Environmental Policy & Planning, 21*(5), 445–463.

Mliless, M., Kerouad, S., & Yachoulti, M. (2021). Environmental education and environmental humanities bridging the gap between teaching and research. In M. Larouz, S. Kerouad, M. Mliless, & H. Laabidi (Eds.), *Current issues in environmental discourse* (pp. 40–64). Moulay Ismail U Pub. ISBN: 978-9920-620-02-4

Mühlhäusler, P., & Fill, A. (2001). *The ecolinguistics reader: Language, ecology and environment.* Continuum.

Ponton, D. M. (2024). Language and ecology in social imaginaries: Ecolinguistic perspectives. *Journal of World Languages, 10*(2), 273–279. https://doi.org/10.1515/jwl-2024-0052

Radovanović, A., & Pešić, D. (2024). Ecolinguistics and tourism. In D. Cvijanović (Ed.), *Tourism and green investments (TISC 2024): Conference proceedings* (pp. 54–63). Retrieved from: https://doi.org/10.52370/TISC2454AR

Ramos, C. M., De Almeida, C. R., & Fernandes, P. O. (2019). *Handbook of research on social media applications for the tourism and hospitality sector*. Business Science Reference.

Reisinger, Y. (2015). *Transformational tourism: Host perspectives*. CABI.

Roudies, N. (2013). Vision 2020 for tourism in Morocco: Focus on sustainability and ecotourism. In *Expert group meeting on ecotourism, poverty reduction & environmental protection*. Retrieved from: https://sustainabledevelopment.un.org/content/documents/4104roudies.pdf.

Rudd, J. A., Horry, R., & Skains, R. L. (2020). You and CO_2: A public engagement study to engage secondary school students with the issue of climate change. *Journal of Science Education and Technology, 29*, 230–241. https://doi.org/10.1007/s10956-019-09808-5

Sarfaraz, H. S., Sedaghat, M., Maknoon, R., & Zavadskas, E. K. (2015). Sustainable tourism: a comprehensive literature review on frameworks and applications. *Economic Research-Ekonomska Istraživanja, 28*(1), 1–30. https://doi.org/10.1080/1331677X.2014.995895

Selkani, I. (2020a). Management of protected areas and ecotourism, what relationship? *Austin Journal of Business Administration and Management, 4*(1), 1–5. www.austinpublishinggroup.com

Selkani, I. (2020b). Ecotourism, travelling, and discovering nature consciously: Case of Atlas Kasbah – Agadir – Morocco. In C. Brătianu, A. Zbuchea, F. Anghel, & B. Hrib (Eds.), *STRATEGICA 2020 – Preparing for tomorrow today* (pp: 301–311). https://strategica-conference.ro/strategica-2020/

Sofronov, B. (2017). Impact of sustainable tourism in the travel industry. *Annals of Spiru Haret University Economic Series*, 4, 85–94.

Stibbe, A. (2015). *Ecolinguistics: Language, ecology and the stories we live by*. Routledge.

Stibbe, A. (2017). Positive discourse analysis: Rethinking human ecological relationships. In A. Fill & H. Penz (Eds.), *The Routledge handbook of ecolinguistics* (pp. 165–178). Routledge.

Tureac, C. E. (2008). Types and forms of tourism. *Acta Universitatis Danubius, 1*, 92–103.

Vellas, F., & Mehadji, Z. (2005). Les stratégies touristiques du secteur privé au Maroc [The tourism strategies of the private sector in Morocco]. *Téoros, 24*(1), 31–36.

Zhang, R. (2022). The year's work in ecolinguistics 2021. *Journal of World Languages, 8*(1), 141–163. https://doi.org/10.1515/jwl-2022-0009

7

Colonial thought, eco-industrialism and speciesism in the discourse of Argentina's food production control agency

Diego L. Forte

Introduction

Public policies constitute a particular form of practice that combines a discursive dimension with the material enforcement, with direct impact on the social dimension. Both for the Weberian visions of the state and for those that follow the theory of organizations, the discourses related to the different areas of a government align themselves, from their own specificity, with the objectives proposed by the head of the national government, establishing the unity of management (Muller, 2002). In this framework, the discourse of the National Agri-Food Health and Quality Service (SENASA for its initials in Spanish: Servicio Nacional de Sanidad y Calidad Agroalimentaria) is in line – to a certain extent – with that of the nation's presidency. However, its particular specificity leads it to relations with other discourses and other representations in the first place, which are, at a later stage, reconciled with the national perspective.

SENASA is the official food production control agency in Argentina. According to the Argentine government website, its main duties are to 'plan, organize and execute specific programs and plans that regulate production, directing it towards obtaining safe food for human and animal consumption'.[1] Nevertheless, discourses produced by the agency for the public knowledge introduce elements that conflict with that affirmation.

In previous research, we pointed out that coloniality is at the root of Argentina's national identity, and it is materialized through different kind of texts

(Forte, 2019). In this context, we believe that SENASA's discourses are based in a neoclassical economics narrative that articulates colonial, eco-industrial and speciesist meanings (Anshelm & Hultman, 2015; Hribal, 2007). Following this affirmation, the present work aims to analyse the construction of this narrative in two audiovisuals issued by SENASA for the general public. Our analysis adopts ecolinguistics as main theoretical framework for econarratives (Stibbe, 2015), and Lim and Tan-Chia (2023) for multimodal transposition. We also take the analytical tools presented by Kress and van Leeuwen (2005) for image analysis, Ekman and Friesen (1975) for facial expression, and Forte (2023) for music analysis.

Our corpus is composed by two audiovisuals taken from the section 'el SENASA en los medios' [SENASA in the media] on the agency website. This section is dedicated to disseminating the agency's work to the general public, and thus the texts are elaborated for non-specialists.

Thought and enforcement: The Argentine way

Colonial thought, neoclassical economics and speciesism in Argentina

National identity discourse in Argentina manifested colonial features from its origins. After the declaration of independence from the Spanish crown in 1810, discourses produced in the country started to build a national character by adopting European models. Literature and politics were the first genres that established the vision that helped set the foundations of the national identity.

The first descriptions of the local landscape presented derogatory visions and appeared in the writings of English travellers. According to Prieto (1996), during the decades of 1820 and 1830, large numbers of travellers crossed the Pampas and the Andes from Argentina to Chile, Bolivia and Peru in search of mining opportunities for British companies and narrated their experiences in different books published after their adventures in the region. The writings of John Miers (1826), Alexander Caldcleugh (1825), Francis Bond Head (1826), Joseph Andrews (1827), John Beaumont (1828) among others constitute, following Prieto, the first step towards an Europeanizing identity. These travellers depicted the Pampas as a desert, a plain that never ends, and compared them to European territories, a very different landscape. Therefore, comparisons were mostly negative regarding the local geography, highlighting the lack of civilization in the South American territories and the need to implement a European way of exploitation for them.

The first generation of Argentine intellectuals – known as Generation of 37 due to the date of the first literary salon in Buenos Aires – continued these ideas, establishing the dichotomy that drove the ideological development of the country in the following century: civilization or barbarism. In the eyes of these intellectuals, the wilderness of South American territories – and their native inhabitants – had to be civilized through European thought. This is openly manifest in the writings of Juan Bautista Alberdi, one of the ideologues of the Argentine Republic and author of *Bases y puntos de partida para la organización nacional* [Groundings and Starting Points for National Organization]. In this work, Alberdi states:

Original Text

¿Queremos plantar y aclimatar en América la libertad inglesa, la cultura francesa, la laboriosidad del hombre de Europa y de Estados Unidos? Traigamos pedazos vivos de ellas en las costumbres de sus habitantes y radiquémoslas aquí. ¿Queremos que los hábitos de orden, de disciplina y de industria prevalezcan en nuestra América? Llenémosla de gente que posea hondamente sos hábitos. Ellos son comunicativos; al lado del industrial europeo pronto se forma el industrial americano. La planta de la civilización no se propaga de semilla. Escomo la viña, prende de gajo. (Alberdi, 1852, p. 97)

English Version

Do we want to plant and acclimatize in America the English freedom, the French culture, the industriousness of the man of Europe and the United States? Let's bring living pieces of them in the customs of its inhabitants and settle them here. Do we want the habits of order, discipline and industry prevail in our America? Let's fill it with people who deeply possess those habits. They are communicative; next to the European industrialist is soon formed the [South] American. The plant of civilization does not propagate from seed. It's like the vineyard; it takes hold. (Author's translation)

This construction of the nation prevails in foundational national literary works such as *La cautiva* [The Captive] by Esteban Echeverría, and *Facundo* by Domingo Faustino Sarmiento. Regarding the Pampas, Sarmiento goes even further:

Original Text

El mal que aqueja a la República Argentina es la extensión: el desierto la rodea por todas partes, y se le insinúa en las entrañas; la soledad, el despoblado sin una habitación humana, son, por lo general, los límites incuestionables entre unas y otras provincias. Allí, la inmensidad por todas partes: inmensa la llanura, inmensos los bosques, inmensos los ríos, el horizonte siempre incierto, siempre confundiéndose con la tierra, entre celajes y vapores tenues, que no dejan, en la lejana perspectiva, señalar el punto en que el mundo acaba y principia el cielo. Al sur y al norte, acéchenla los salvajes, que aguardan las noches de luna para caer, cual enjambres de hienas, sobre los ganados que pacen en los campos y sobre las indefensas poblaciones. (Sarmiento, 1845, p. 67)

English Version

The evil that afflicts the Argentine Republic is its extension: the desert surrounds it on all sides, and it insinuates itself into its bowels; solitude, depopulation without human habitation, are, in general, the unquestionable limits between one province and another. There, the immensity everywhere: the plain is immense, the forests immense, the rivers immense, the horizon always uncertain, always merging with the earth, between clouds and faint vapours, which do not allow, in the distant perspective, to point out the place where the world ends and heaven begins. To the south and north, the savages stalk it, waiting for the moonlit nights to fall, like swarms of hyenas, on the cattle that graze in the fields and on the defenceless populations. (Author's translation).

This thought became the starting point for the use of livestock as a way to populate the desert (Antagonism & Practical History, 2022, p. 22) in order to eradicate the void and bring civilization into it. From then on, beef and Argentina became synonyms.

In the last decades of the nineteenth century, Argentina became one of the main producers of the foods consumed by the industrialized economies of Europe, and meat played an important role in that context (Rayes, 2015). Although beef was not the most exported product, its importance is linked to the introduction of foreign capital in the conformation and expansion of the industry, and the dependence on Great Britain in commercial terms (2015, p. 9). This growth as a meat provider combined the colonial features of the Argentine identity with capitalist traits to structure the backbone of the current dominant discourse of national identity in the country. In it, coloniality and neoclassical economics are two faces of the same coin that alternate as organizers of meaning.

Discursive enforcement of colonial speciesism

Although regulations and law enforcement regarding food production in Argentina have a long tradition, the SENASA as such is a relatively new agency. The functions of what is today called the National Agri-Food Health and Quality Service began at the end of the nineteenth century, basically to provide health guarantees for exports of raw materials. According to the information provided by its website, the first tasks of the agency were focused exclusively on imports and exports. At later stages of the development of the country, new structures were incorporated to involve internal market controls.

During the nineteenth century until the end of the 1960s, the enforcement of alimentary regulations was carried out by the Ministry of Agriculture with the help of the federal security forces. In 1972, the National Service for Animal Health (SENASA) was created for animal origin food control, and two decades later, the Argentine Institute for Vegetal Health and Quality (IASCAV for its initials in Spanish, *Instituto Argentino de Sanidad y Calidad Vegetal*) for vegetal origin foods. Both agencies were merged in 1995 in the current SENASA.

As official control agency, SENASA is the authority for enforcing alimentary regulations. Therefore, we can say it is the regulator of social practices involving non-humans. The following sections analyse two multimodal texts published in the section 'SENASA in the media' of the official webpage of the agency. This section collects appearances in the press by agency officials, and statements and publications related to food safety issued by the agency itself. The first text analysed here is a publication made by the agency on the national government page (www.argentina.gob.ar) to announce the opening of the Chinese market for Argentine fishing products. This is a complex text that involves plain text, an audiovisual file and a photo. The second text is a press release issued by SENASA describing the situation posed by the bird flu outbreak and the measures taken by the agency to prevent its entry into the country.

Critical ecolinguistics and multimodality

As a transversal discipline, ecolinguistics provides a critical look at representations – the mental models – and allows for deconstructing the power relations among humans, non-humans and ecosystems built through discourse.

> Ecolinguistics can explore the more general patterns of language that influence how people both think about, and treat, the world. It can investigate the stories we live by – mental models that influence behaviour and lie at the heart of the ecological challenges we are facing. There are certain key stories about economic growth, about technological progress, about nature as an object to be used or conquered, about profit and success.
>
> (Stibbe, 2015, p. 2)

This theoretical deconstruction evidences how particular narratives structure meanings around specific axes and guide particular readings. The capitalist economic discourse, for example, organizes meanings around economic concepts, establishing a logic that rules other discourses such as those of scientific dissemination, constructing narratives that, for instance, describe science from the idea of economic development. In this way, public policies are seen as tools at the service of commerce which, in turn, is seen as a form of the development of the country. Thus, humans, non-humans and everything included in the ecosystem are transformed into tradable resource within a capitalist logic that places the country as a supplier of raw materials subordinated to industrialized countries.

On the other hand, multimodal approaches evidence the structuring of semiotic resources in audiovisual material to describe how they build meaning and orient particular readings. We think about meaning as a transposition among modes (Lim & Tan-Chia, 2023, p. 92) in which every stage contributes with particular elements to a global multimodal construction.

In the global multimodal construction, one mode takes the lead – usually speech or verbal written text – establishing a gradation of modes that subordinate the remaining modes to the leading or dominant.

In this construction, emotional manifestations play a fundamental role: they frame the elements in particular moods to generate empathy with the reader/observer/audience, provoking some type of attachment or more direct relationship between the observer and the message. As humans living in society, we act based on the emotional manifestations of those around us. These manifestations guide, to a greater or lesser extent, our responses:

> Reflecting on and acting upon the emotions of others is a basis of empathy, and the inability to read emotions accurately is likely to be associated with less empathy. With less empathy, there is less kindness and consideration for fellow humans. Moral laws and implicit ethical codes are loosened. The nature of friendships changes, as does the nature of collegiality. Under a veil of

institutionalized and ritualized politeness can be an underlying current of cold interpersonal attitudes. (Matsumoto, 2007, p. 47)

So, emotional manifestations present in material related to public policies – a tough subject for a wide audience – are aimed to raise at least a small connection with the information provided. These entire resources construct a narrative that reflects the combination of coloniality and neoclassical economics structuring speciesism in Argentina's public policies.

The first multimodal text that we analyse below is a publication issued by SENASA's press office to communicate the opening of the Chinese market for Argentine trout. It includes written text, a photo and a video, and was shot by the office and edited inside the agency. The second text is a press release issued by SENASA describing the situation posed by the bird flu outbreak and the measures taken by the agency to prevent its entry into the country.

China abrió su mercado al ingreso de trucha arco iris cultivada en agua dulce en Patagonia [China opened its market to the entry of rainbow trout grown in freshwater in Patagonia]

This article was published in the webpage of the government of Argentina on 5 January 2024 as an institutional post, with no explicit author assigned. It is formed by a small written text, a 3:15 minute video and illustrated by a photo. The written part operates as frame for the video that constitutes the main attraction of the article. Following the headlines, the video is the first item, then the small text comes right after, and the photo is in the middle of it. The words highlight passages of what is said in the video by Fabián Ballesteros, SENASA's Director of Safety in Fishing Products and the protagonist in the images. The photo illustrates the work involved; it shows a worker raising some trout with a landing net.

The story built by the combination of multimodal resources is that of commercial success. The written text, the photo and the elements deployed in the video serve a narrative about a team effort that ends successfully: SENASA teams up with the company that breeds the trout to get permits from the Chinese sanitary authority to export. But the name of the company is erased, neither mentioned in the written text nor the video. Therefore, this is a commercial success for an official sanitary authority only and for the Argentine fishing industry as a whole.

The written text and the images – of the video and the photo – are the first things an observer can see when entering the website. Although the layout first presents the video, given the amount of information it is unavoidable to read the words first: they are just a few lines and seem to give some advance information of what is in the video.

A first look at the thematic structure of the written text gives a clue about which participants and processes are involved: the responsibility of SENASA in actional processes is mitigated. Permits were granted by the Chinese authorities:

Original Text	**English Version**
La República Popular de China autorizó el ingreso de trucha arco iris (Oncorhynchus mykiss) congelada, proveniente de la cría de acuicultura en agua dulce de la Argentina.	The People's Republic of China authorized the entry of frozen rainbow trout (Oncorhynchus mykiss), from freshwater aquaculture farming in Argentina. (Author's translation)

From a systemic-functional perspective, SENASA's role appears in a rhematic position, that is, remarking on the efforts and the achievement but also mitigating the involvement of the agency:

Original Text	**English Version**
Las gestiones para lograr esa apertura y establecer requisitos de sanidad e inocuidad estuvieron a cargo del Servicio Nacional de Sanidad y Calidad Agroalimentaria (SENASA), organismo descentralizado de la Secretaría de Agricultura, Ganadería y Pesca de la Nación y de la Administración General de Aduanas de China (GACC).	The efforts to achieve this opening and establish health and safety requirements were carried out by the National Agri-Food Health and Quality Service (SENASA), a decentralized body of the Ministry of Agriculture, Livestock and Fisheries of the Nation and the General Administration of Customs of China (GACC). (Author's translation)

Only when Ballesteros's words are quoted, responsibility (linguistic agency) appears in full presence. The text builds the idea that it is not SENASA but its officials that have an active role in the processes that manifest rather visually than verbally:

Original Text

El director de Inocuidad de Pesca y Acuicultura del Senasa, Fabián Ballesteros, detalló que el cultivo de truchas 'es un tipo de producción que se está desarrollando en los últimos dos años al que el SENASA está acompañando en la gestión para la apertura de mercados internacionales y en el acceso al mercado interno'.

English Version

The director of Fisheries and Aquaculture Safety at Senasa, Fabián Ballesteros, explained that trout farming 'is a type of production that has been developing in the last two years that SENASA is accompanying in the management for the opening of international markets and in access to the internal market'. (Author's translation)

Nevertheless, when Ballesteros describes his work in the first person, he uses pronouns corresponding to the first plural person: 'we' (SENASA):

Original Text

Estamos trabajando hace tiempo con una empresa que produce truchas
Acompañamos mercados internacionales, productos
Generamos un vínculo de consulta con las autoridades sanitarias y bloques comerciales de los países de destino

Ofrecemos garantías para que el producto pueda acceder al mercado

En ese desarrollo de gestión hemos abierto el mercado de la UE para productos de la acuicultura y especialmente para trucha 'panzai'

Continuamos trabajando para abrir el mercado paratrucha fresca

English Version

We have been working for a long time with a company that produces trout
We accompany international markets
We generate a query link between sanitary authorities and commercial blocks

We offer guarantees so that the product can access the market
In this management development we have opened the EU market for aquaculture products and especially for panzai trout
We continue working to open the market for fresh trout

His words minimize health requirements, the official function of SENASA, reducing it to paperwork:

Original Text	**English Version**
Resta dar cumplimiento a un procedimiento que tiene el GACC (autoridad sanitaria China) para completar unos formularios y cuestionarios que así lo requiere la autoridad china, para poder finalmente exporter trucha congelada	It remains to comply with a procedure that the GACC (Chinese health authority) has to complete some forms and questionnaires required by the Chinese authority, in order to finally be able to export frozen trout

He also appeals to some universalizations in building the role of China as a big market because of its sanitary requirements and the volume of products they consume:

Original Text	**English Version**
Todo el mundo valora el mercado de China por el volumen de consumidores y por la calidad que también exige	Everyone values the Chinese market for the volume of consumers and the quality it also demands
Es muy importante la apertura de este mercado para este producto	The opening of this market for this product is very important

This universalization establishes the importance of China and the EU as something known by the whole world, therefore positioning Argentina as a prestigious place just because it is doing business with them.

There is no mention, either on the main page or in the video, about environmental policies, risks or measures. Trout is raised in artificial pools built in valleys in Patagonia with a potentially significant impact on the environment (Jeanson et al., 2022; Dell, 2023). Nevertheless, this is not mentioned.

The audiovisual material is structured through the voice of Fabián Ballesteros, who speaks during the whole running time. In the background, a sort of institutional music can be heard at a very low volume. It is music used by SENASA for all the audiovisual material made by and for the agency: instrumental, up-tempo average TV music in major tonality. This audio track (voice + music) is illustrated by images of Ballesteros being interviewed that alternate with tales of the fish, workers in the processing plant cutting the fish, and the valley where the pools and the plant are located.

The ideational metafunction shows actional processes that become – through Ballesteros's descriptions – conceptual ones. The narrative constructed by his voice grants the images the role of illustrations. Following this logic, the fish are not 'doing' (in this case, swimming) but 'being' fish and doing what fish do. Fish swimming in circles inside the pools could be considered an actional process carried out by the animals, considered as individuals with agency of their own and behaving strangely due to their confinement. But words at the bottom of the screen read 'International commerce: Patagonia trout exports, EU and China markets open'. These words, along with Ballesteros's speech, orient meaning towards a representation of the trout as material and not as individuals. A similar case can be considered for human processes (Figure 7.1). In several parts of the audiovisual, we can see the work inside the processing plant and the operatives removing thorns from fish carcasses. This can be thought as an actional process developed by the operatives that has an impact on the fish. Nevertheless, these images too are subjected to Ballesteros's voiceover and also become a conceptual example of what is said in the audio track.

In this case, workers are neither 'doing' their jobs but rather 'being workers' and doing what workers do. Only when Ballesteros's face is shown can we, as observers, see who is speaking and, therefore, consider the image as an actual actional verbal process with a known and defined actor.

The analysis of the interpersonal metafunction evidences that images build an act of offer: there is no participant looking directly at the camera, not even Ballesteros, who is looking to the right where the interviewer seems to be – he is out of shot, and we can only see his arm holding a microphone. Workers are

Figure 7.1 Trout and operatives involved in actional-conceptual processes.

focused on their tasks, and fish are not shot at a close range to see their faces. The only face that can clearly be seen is Ballesteros's, taken in a short medium shot. His face evidences a neutral position of its features: just a few rhythmic movements to mark parts of his speech, but no emotional expression is shown in the whole video. The only openly manifested emotional marker in the video is music that, nevertheless, is played at a very low volume (Forte, 2023). Music adds a positive attitude towards the content of the other modes through major tonality and the up-tempo structure, also providing a sense of dynamics that makes the video pass quickly and smoothly.

On the other hand, the size of the shot divides images that function as examples from those that underpin the narrative, building a sort of gradualism between what is narrated and the narrator (Ballesteros). Takes of the fish are long shots; workers are shown in long or full-size shots, while Ballesteros is shown in medium shot (Figure 7.2). Thus, as we mentioned, the voice organizes the visual disposition, granting a central role to the speaker's face, while fish and workers are presented as surrounding this central place where Ballesteros is (Figure 7.3).

The angle of the shot is high only for the fish in the pool. The other participants are shot at eye level (Figure 7.4).

These takes construct relative power in terms of equality among humans – between every other participant represented, and between the represented and the interactive participants – but this is not the case regarding the fishes.

Finally, during the entire audiovisual, images are narrativized by small portions of text at the bottom of the screen that explain the context of what is

Figure 7.2 Ballesteros's verbal process.

Figure 7.3 Fish and operatives in long shot.

Figure 7.4 High angle on fish, eye level on humans.

shown through the images: 'trout production', 'SENASA guaranties . . .' and so on, orienting the interpretation of the voice and images.

The textual metafunction divides resources to build the system of given/new: while the given is constituted by the fish, and the workers and the officials of SENASA, the new is constructed first by the voice narrating the way the EU and Chinese markets have been opened, and then by the words on the screen that anchor the images as belonging to the 'opening of the markets' topic.

Outside the video and back on the main page, the photo that illustrates the text shows a worker taking fishes out of the water with a landing net. In the analysis of the ideational metafunction, this can be thought of as an actional process: fishes are the affected participant. But if we follow the logic presented through the video and consider the speciesist perspective, the fish are just material. In that case, the process does not have affected participants (Figure 7.5).

Figure 7.5 Actional process.

The interpersonal metafunction evidences an act of offer and an upper angle shot that builds the represented participant as small, with less relative power in relation with the viewer. However, this image seems to try to depict the material effort involved in the opening of the markets by presenting a representative of the company that is not the owner or the president but a worker, reinforcing in this way the meaning that 'success is achieved through hard team work'.

The textual metafunction shows the human participant in the centre of the image and the fish as peripheral: they are almost impossible to identify as individuals.

Coloniality features appear through the words and the video along with the commercial success story of teamwork. Although in a subtle way, a few elements appear repeatedly in different official constructions and build the following ideas:

1. Prestigious markets pose important demands: China y Unión Europea son dos mercados altamente exigentes y con un volume de consume muy importante [China and the European Union are two highly demanding markets with a very significant volume of consumption]; Todo el mundo valora el mercado de China por el volumen de consumidores y por la calidad que también exige [Everyone values the Chinese market for the volume of consumers and the quality it also demands].
2. We are below those demands and must team up to achieve the necessary level. They – prestigious markets – are the beacon to follow: the EU is mentioned, although the news is that the Chinese market has opened.

The narrative constructed in this article through the combination of text and audiovisual materials poses a narrative in which commercial and sanitary success go hand in hand, while the ecosystem and species other than human are considered as necessary resources. In ecolinguistic terms, this narrative is based in an anthropocentric destructive ecosophy (Stibbe, 2024, p. 33) that also evidences traits of coloniality, praising EU – and also Chinese – markets as highly valuable due to international prestige.

Influenza aviar: recomendaciones para quienes tengan aves de traspatio [Avian flu: recommendations for backyard bird's breeders]

This audiovisual was published in SENASA's YouTube channel on 3 February 2023. In it, a voiceover describes the production locations in Argentina for avian meat, foreign markets, sanitary conditions and how to avoid risks in backyard birds through the international outbreak of bird flu that took place during that year. The voice-off narrative is interrupted by a short interview of Jorge Dillon, then president of the agency, describing the measures taken to avoid the entering of the disease into the country. In the video, images take the lead: the topic discussed is illustrated by videos of different birds, farms and some SENASA personnel working in the field, as well as chicken images. The story narrated through the video is that of the health of foods being a fundamental feature for commercial success and the risk bird flu presents for human health and markets.

Original Text	**English Version**
El servicio nacional de sanidad y calidad agroalimentaria SENASA que ejecuta un constante trabajo preventivo con controles periódicos que estudian anualmente a las aves comerciales las de razas puras las no comerciales de producción familiar y las aves silvestres constatando la ausencia de casos de aves infectadas con él de la gripe aviar	The national service of health and agri-food quality SENASA, which carries out constant preventive work with periodic controls that annually s tudy commercial birds, pure breeds, non-commercial family-produced birds and wild birds, confirming the absence of cases of birds infected with avian flu

Birds are, bluntly, considered 'commercial' from the beginning and, therefore, are the main element at risk in the context of the international outbreak. The relation between sanitary measures in food and commercial interests is remarked upon several times, in addition to the prevalence of commercial interests over bird life, naturalizing the carelessness of producers in a disease of low danger to humans:

Original Text	**English Version**
Nuestro servicio cada tanto hace algún entrenamiento algún ejercicio de simulación del brote tiene personal y técnicos capacitados	Our service occasionally conducts some training and some simulation exercises of the outbreak, and has trained personnel and technicians
Los productores la manera más aconsejable de prevenir el virus es la práctica de la bioseguridad por ejemplo asegurar las puertas y los galpones donde se mantienen las aves permitir solamente la entrada de las personas que trabajan en la granja	For producers, the most advisable way to prevent the virus is to practice biosecurity for example, securing the doors and sheds where the birds are kept and allowing only the entry of people who work on the farm

Given the low risk the flu poses for humans, the main problem is a commercial one. Environmental conditions and manipulations enacted by humans that made the propagation of the disease possible are not mentioned. Human responsibility is minimized, and the narrative focuses on the measures to avoid the flu. The most important effect mentioned in the discussion is commercial:

Original Text	**English Version**
La limpieza y desinfección a los trabajadores entre otras teniendo en cuenta la situación mundial es necesario extremar las medidas de prevención y de alerta a nivel país para evitar su posible ingreso al territorio nacional y mantener a la argentina libre de este flagelo y su impacto comercial	Cleaning and disinfection of workers, among others, taking into account the global situation, it is necessary to take extreme prevention and alert measures at the country level to avoid its possible entry into the national territory and keep Argentina free of this scourge and its commercial impact

The role of birds is reinforced when they are included in the axis of tradition, naming them as *aves de traspatio* (backyard birds). The images show different processes in different parts of the verbal discussion.

The ideational metafunction constructs, in the first seconds, a combination of the voiceover presenting the national and international context for the outbreak with clean images of the birds in pens in large and close shots. No filth or sick birds are shown – something very different from the birds raised for food exploitation. In these images, the birds become participants in conceptual processes. They appear as representatives of a class, 'the birds'. This establishes that birds are for human consumption: any bird can contract the flu; nevertheless, the only concern raised by the agency is towards those considered as 'commercial'. When human participants appear, they do so in other parts of the video, not sharing space with the non-humans involved in the verbal conversation. When birds are on screen, the verbal process is carried out by the voiceover and the text printed on the screen ('bird flu grows in the world'). In this case, there is an actional verbal process between the voiceover and the listener, and the crowded birds are shown as a token of the farms, as a conceptual process.

In the interpersonal metafunction, the face of Jorge Dillon with a neutral emotional face constitutes one of the main connections with the observer. Dillon does not address the camera; he talks with the journalist who is interviewing him outside the frame. This configures an act of offer: there is no asking for involvement in the image but rather providing information. The expression on his face constructs a serious offer: the subject at hand calls for seriousness, although he is not addressing the public directly. On the other hand, non-humans are shown in an offer act, and their faces cannot be seen in the crowded shelves. Human faces are taken in close shots, and long shots are used for the chickens except for a few takes where groups of four or five birds are shown in close shots. The images of the overcrowding of the birds in pens, in which the breeding hens are placed, transform them into a collective subject, denying their individuality even when they are shot closely by the camera.

The camera angle is at eye-level for both humans and chickens, establishing a relation of relative power between the observer and the represented participants – being these humans or non-humans – as equal. This is only reversed by the crowded situation in which the birds are shown: as we mentioned above, birds are constructed as a collective individual, while Jorge Dillon, the only human in the image, is perfectly individualized (Figures 7.6 and 7.7).

Figure 7.6 Verbal and conceptual processes simultaneously.

Figure 7.7 Long shoot for chickens.

Figure 7.8 Given/new system.

Figure 7.9 Healthy birds and SENASA's measures.

The textual metafunction builds the system of given/new using the texts in the frames. When birds are shown in the centre of the image – the given – the bottom text reads, 'Biosafety in farms is the only wayto prevent the disease' (Figure 7.8).

In this way, hens have prominence in the big picture while the text introduces a new element to interpret their appearance. The colonial element appears timidly again in the expression of Jorge Dillon:

Original Text	**English Version**
Una enfermedad que sería bastante complicada para nosotros y que podría generar una crisis bastante importante (en términos comerciales, sobre todo con Europa)	A disease that would be quite complicated for us and that could generate a fairly important crisis (in commercial terms, mostly with Europe)

This is affirmed and justified by Dillon in the later description of the measures SENASA is taking to minimize the impact of bird flu in international commerce. At the moment this topic is discussed, images of healthy birds are shown, constructing the idea that the work of the agency allows good production and avoids problems with big foreign markets (Figure 7.9).

Conclusion and summary

Although public policies arise from particular ecosophies, they constitute a way to enforce them. Considered as narratives, or part of bigger stories, they establish legal boundaries for ethics and discursive relations with the world around us. In this sense, they become a strong instrument to regulate material actions, attitudes and emotions towards other species and the ecosystem.

As a discursive genre, public policies combine meanings from other discourses such as coloniality and neoclassical economics, although they are not always openly manifest in the texts. They rather conform to a net of signs that can represent different ideologies, but so organized they are part of a complete narrative and, therefore, part of its global meaning.

Argentina's national narrative constructs a colonial perspective that has organized discursive meanings to this day. This narrative describes the country as an agro-exporter nation that can only be successful by exploiting other species and selling its production to Europe and other countries. In this story, non-human agency is erased, and instrumentalization with commercial means is naturalized. Environmental issues are minimized along with health issues, and ecological implications are not even mentioned.

The focus in both multimodal texts is on trade and intensive production as a synonym for development and success, without considering environmental impacts, cruelty to animals or diseases to humans. According to this narrative, commercial success justifies the means of production. Furthermore, this commercial success is provided by the reputation of the clients: exporting from Argentina means exporting to the EU, which is why every other destination is compared to Europe to establish its importance, even when, in commercial terms, other destinations generate greater profits for the country, as is the case with China. The Argentine Europeanizing colonial identity always follows the European ideal.

In health terms, narratives created by authorities should refer to safety standards related to health, safety and well-being rather than commercial variables. This situation concerns us not only as linguists and discourse analysts but also as citizens committed to our ecological and natural social context, in the context of the current climate crisis. Analysing what our elected authorities are telling us about the discursive world surrounding us becomes imperative.

Note

1 www.argentina.gob.ar/senasa

References

Alberdi, J. B. ([1852] 2017). *Bases y puntos de partida para la organización política de la República Argentina*. Biblioteca del Congreso de la Nación.

Andrews, J. (1827). *Journey from Buenos Aires, through the provinces of Cordova, Tucuman, and Salta, to Potosí*. John Murray.

Anshelm, J., & Hultman, M. (2015). *Discourses of global climate change: Apocalyptic framing and political antagonisms*. Routledge.

Antagonism & Practical History. ([1999] 2022). *Bestias de carga*. Ochodoscuatro.

Beaumnt, J. (1828). *Travels in Buenos Aires, and the Adjacent Provinces of the river of La Plata with observations*. James Ridgway.

Bond Head, F. (1826). *Rough notes taken during some rapid journeys across the Pampas and among the Andes*. John Murray.

Caldcleugh, A. (1825). *Travels in South America during the years 1819-20-21*. John Murray.

Dell, C. (2023). What fish farming really means for the environment, animals and people. *Sentient Media*. November 8. https://sentientmedia.org/fish-farming/

Ekman, P., & Friesen, W. V. (1975). *Unmasking the face: A guide to recognizing emotions from facial clues*. Prentice-Hall.

Forte, D. (2019). La construcción del discurso identitario nacional argentino: la historia de la carne [The construction of the Argentine national identity discourse: The history of meat]. *Question, 1*(64). https://doi.org/10.24215/16696581e233

Forte, D. (2023). Music and discourse: A systemic-functional approach for music analysis in multimodal contexts. *Multimodality & Society, 3*(1), 69–81. https://doi.org/10.1177/26349795231153963

Hribal, J. (2007). Animals, agency, and class: Writing the history of animals from below. *Human Ecology Review*, 14(1), 101–112.

Jeanson, A., Gotzek, D., Mam, K., Hecht, L., Charvet, P., Eckerström-Liedholm, S., Cooke, S., Pool, T., Elliott, V., & Torres, Y. (2022). Inland fisheries management – Case Studies of Inland Fish. In Thomas Mehner & Klement Tockner (Ed.), *Encyclopedia of inland waters* (Second ed., pp. 343–354). Elsevier. https://doi.org/10.1016/B978-0-12-819166-8.00170-5

Kress, G., & van Leeuwen, T. (2005). *Reading images: The grammar of visual design*. Routledge.

Lim, F. V., & Tan-Chia, L. (2023). *Designing leraning for multimodal literacy*. Routledge.

Matsumoto, D. (2007). Playing catch with emotions. *Journal of Intercultural Communication*, (10), 39–49.

Miers, J. (1826). *Travels in Chile and La Plata*. Baldwin.

Muller, P. (2002). *Las políticas públicas*. Universidad Externado de Colombia.

Prieto, A. (1996). *Los viajeros ingleses y la emergencia de la literatura argentina*. Sud americana.

Rayes, A. (2015). Destinadas a un destino: Los inicios de las exportaciones argentinas de carnes frigoríficas, 1883–1913. *Estudios Interdisciplinarios de América Latina y el Caribe*, *26*(1), 7–30. http://hdl.handle.net/11336/10205

Sarmiento, Domingo, F. ([1845] 1993). *Facundo o Civilización y barbarie. Introducción de Carlos Altamirano.* Bs. As., Espasa Calpe - Colección Austral.

Stibbe, A. (2015). Ecolinguistics. Language, ecology and the stories we live by. Routledge.

Stibbe, A. (2024). Econarrative: Ethics, ecology and the search for new narratives to live by. Bloomsbury.

Corpus

China abrió su mercado al ingreso de trucha arco iris cultivada en agua dulce en Patagonia. (2024). Available online: https://www.argentina.gob.ar/noticias/china-abrio-su-mercado-al-ingreso-de-trucha-arco-iris-cultivada-en-agua-dulce-en-patagonia. Accessed January 17, 2024.

Influenza aviar: recomendaciones para quienes tengan aves de traspatio (2023). Available online: https://www.youtube.com/watch?v=07CggbxizpQ. Accessed January 17, 2024.

8

Ecological identity in Kenya

Philip W. Rudd and Peter Githinji

Introduction

Kuinua Ukuaji Kijani 'Raising Green Growth' begins the theme for the first Africa Climate Summit (ACS) in Nairobi, Kenya in September 2023. To underscore the growing sense of urgency that Africa act against climate change, the president of Kenya, William Ruto, drove himself in a small electric vehicle to meet with the elites of twenty African countries among others (Waweru, 2023). Although awareness of environmental crises proliferates on social media, public information about climate change comes mostly from mainstream news media (Painter, 2013). Mainstream news reports tend to slant in favour of powerful sources and thereby promote a worldview that aligns with the elite establishment (Mautner 2008: 33). Mainstream news media moreover grows readership through instructional use as a stimulus to create interest in reading and as a tool to build a knowledge base for good citizenship (Gardner & Sullivan, 1999: 12). In fact, Nation Media Group (NMG), one of the largest media companies in East and Central Africa, attracting audiences not only in Kenya but also in Rwanda, Tanzania and Uganda, launched the Newspapers in Education Programme (NiE) in 2009, a joint effort with Safaricom, Kenya's largest mobile network operator, to promote a culture of reading and an awareness of public health and environmental issues among school students in Kenya. In this chapter we direct our focus onto one of these powerful elite channels of environmental news. Our analysis examines the frames, lexicon and grammar in influential climate change newspaper stories from one of the most prominent Swahili language newspapers in Kenya. Our focus is on the period 2019 to 2023.

Shaw and Nerlich (2014) have revealed that discourses of prominent international climate governance institutions acting as elite media sources construct unrealistic narratives whose primary objective is amenability to policy making. Atieno and Njoroge (2016) have disclosed that the language of two mainstream Kenyan English newspapers (*The Standard* and *Daily Nation*) in narratives on environmental conflict and tourism caters to human appeal (sensationalism) and marginalizes the environmental angle, thereby becoming destructive to justice and resolution plans. Another investigation (Kitonga et al., 2023) has examined journalistic usage in a Kenyan Swahili gazette (*Taifa Leo*) from what appears to be a rhetorical and literary angle to reveal the paper's goal of selling itself to grow readership. However, no attempt has yet been made to trace the emergence of language and frames in newspaper narratives that are positive or beneficial, and thus conducive to environmental awareness and an ecological identity essential to good climate citizenship.

Two climate conferences significantly bookend the window of time covered in this analysis. The 2019 flank is the 25th session of the Conference of the Parties (COP25) held in Madrid, Spain. It was the final venue of three for the congress after having moved first because climate-change-denying then-president Jair Bolsonaro retracted Brazil's offer to host (Sengupta, 2019). A second move occurred after school students practicing civil disobedience sparked a Chilean everyman's protest against the economic gap between the haves and have-nots in a country long celebrated as prosperous and well-run, making the venue untenable (Londoño, 2019). COP25 is noteworthy among other reasons for its having acknowledged the need to protect the traditional knowledge of indigenous peoples for the benefits to climate action (Dagnet et al., 2019). The 2023 flank is COP28, the 28th annual United Nations (UN) climate meeting. The summit was controversial because its venue was in the United Arab Emirates (UAE), one of ten top oil-producing countries, and the conference president was Sultan al-Jaber, the chief executive of UAE's state-owned oil company. Even more pernicious was the fact that leaked documents showed that the UAE had planned to use its role as host to broker oil deals during the summit in which governments should have been negotiating limits and readying themselves for forthcoming climate changes (Rowlatt, 2023).

In the glaring contrast breaking through the dichotomy of worldviews from COP25 and COP28 summits, we have chosen to focus on news stories that suggest ecological philosophy and good citizenship. Essentially, this chapter endeavours to identify how mainstream news media describes possible responses to climate change and how that language influences the lay reader. Given the importance attributed to the awareness of public health and environmental issues, a better understanding of how newspaper narratives shape Kenyan everyman's framings of climate change can offer guidance in how the process of communication interventions should be targeted. To explore this better environmentally friendly citizenship, we employ ecolinguistics by combining critical discourse analysis (henceforth CDA) with systemic-functional grammar (Halliday & Matthiessen, 2014) to examine how discourse constructs reflect ideologies and create unequal power, and how to place importance on the lives of living beings and social justice. In doing so, we also leverage framing (Cachelin et al., 2010; Lakoff, 2010; Stibbe, 2015) as a tool to grasp how the mainstream media shapes and limits speech, choices and actions of readers and would-be well-informed citizens.

The remainder of this chapter is structured as follows: The methods section provides a general introduction to systemic-functional grammar (Halliday & Matthiessen, 2014), lexicogrammar and transitivity, and other theoretical concepts addressing frames and framing (Stibbe, 2015) and cultural codes (Gavriely-Nuri, 2012). The conceptual framework is then used as a springboard to propose issue frames that may lead to modelling a positive ecological identity. This chapter echoes the work of Venkataraman (2018) in suggesting that diction and word choice are not usually deliberate in the intention to portray a negative or dangerous ecological stance because how 'we speak about the natural world is not a transparent window, because it reflects the culture in which we live and its priorities and values' (Larson, 2011: ix). The final section summarizes the primary findings of the study and suggests future directions for ecological identity research.

Methods

Our methodology for analysing the data was the collection of a range of prototypical texts produced and used by reporters for *Taifa Leo*. In Kenya, this gazette is the only daily Kiswahili (cha Kenya) newspaper, and it holds public reverence for having been around for more than sixty years, having a pedagogical influence on primary and secondary students (Mukhwana & Mukhwana, 2017; Bylanes, 2021; Kitonga et al., 2023). Besides its assertion that readers may *ng'amua ukweli wa mambo* 'understand the truth of the matter', *Taifa Leo* provides a window into the dominant and elite perspectives of Nation Media Group (NMG), one of the largest media companies in East and Central Africa, attracting audiences not only in Kenya but also in Rwanda, Tanzania and Uganda. *Taifa Leo* was also chosen as a site of investigation because of its unique position as Kenya's longest-running indigenous voice of African language. This focus distinguishes *Taifa Leo* from *Mwananchi*, also controlled by NMG but which is Tanzania's leading newspaper, because of its tendency to reflect the perspectives and language of Kenyans (*Kiswahili cha Wakenya*), a fact that disqualifies it as 'suitable', according to Mukhwana and Mukhwana (2017: 61), 'as an instructional tool'. Nonetheless, our motive is to investigate language as it is, not to pursue prescriptive language ideology.

The dataset for the study was initially composed of twenty-seven newspaper articles from *Taifa Leo Dijitali* from 2019 to 2023. The sample set was accessed and compiled through the search feature on the *Taifa Leo* website using the key phrase *mabadiliko ya tabianchi* 'climate change'. Originally, all manner of articles from news reports, editorials, features, opinion pieces and letters to the editor were included. Though the samples could hardly be fully representative of all writing produced by newspaper reporters, they do expose discourses that were prominently employed by a subset of reports during the time studied. Still, a few were eliminated after review to reduce redundancy and to facilitate a more manageable packaging of frames. A final selection of ten items was labelled from A1 to A10 (see Table 8.1) and contains a mixture of news articles for a small, purposeful dataset covering nearly half a decade. Each article was analysed for clause processes and for framing.

To reveal patterns of how the language was used within and across the texts, we focused on linguistic features which combine together to tell narratives about the environment. Our initial effort to see how reporters represented climate change was to examine the transitivity of clauses composed. Transitivity, as described by Halliday and Matthiessen (2014), breaks clauses into types of processes and kinds of nouns or participants represented. Nouns referring to

Table 8.1 Distribution of Process Types

Article	Material	Mental	Relational	Verbal	Existential	Total	Percentage
A1	36	27	27	13	0	103	21
A2	13	0	2	0	0	15	3
A3	80	15	14	11	1	121	24
A4	10	0	1	2	0	13	3
A5	13	4	5	4	2	28	6
A6	29	8	9	15	9	70	14
A7	18	10	8	11	3	50	10
A8	7	13	8	1	1	30	6
A9	6	3	1	1	0	11	2
A10	30	11	4	9	3	57	11
Total	242	91	79	67	19	498	100
Percentage	49%	18%	16%	13%	4%	---	100%

participants in the articles were classified in alignment with the processes of the lexical verbs, which categorize as Existential, Relational, Material, Mental and Verbal. Existential process verbs describe the existence of one participant, an existent. Relational process verbs represent the state of a token and a value. Material process verbs describe action with the cause (Actor). At times there is an object (Affected). Sometimes a second affected occurs (Recipient). Mental process verbs represent emotion, perception or thought of an explorer and an experience. Verbal process verbs present an act of saying or writing (Verbiage) between a Sayer and a Receiver. As is evident, transitivity indicates the kinds of processes that participants are involved in, and specifically whether they perform Actor or Senser or more passively as Affected or Phenomenon. Minor clause processes may vary, and in our analysis, material clauses subsumed behavioural clauses as a subtype; existential clauses sometimes shared a nebulous border with possessive and/or locative relational clauses (See footnote one of Halliday & Matthiessen, 2014: 214).

Critical discourse analysis (CDA) empowered us to identify what was 'between the lines' in texts of environmental narratives (Venkataraman, 2020; Stibbe, 2015; van Dijk, 2011; Martin & Rose, 2003). In CDA for ecolinguistics, framing was used to grasp how the media shaped and limited the language. This ecological framing of environmental issues (Stibbe, 2015; Lakoff, 2010; Cachelin et al., 2010) can have a negative or positive slant. Following Stibbe (2015: 31), we diagnosed texts as either destructive (i.e. ecologically adverse), ambivalent (ambiguous) or beneficial (aligned) to the aims of ecolinguistics. That way, we were able to enumerate issues that needed to be resolved. The purpose then was to be able to recommend that destructive discourses be resisted via our helping to raise awareness of their danger, that ambivalent language be modified constructively to build responsibility, and that beneficial discourses be celebrated and promoted. In this manner, we can encourage social change (Stibbe 2015: 40) that aligns with justice for the environment and the Kenya everyman.

Our analysis of the framing of *mabadiliko ya tabianchi* targeted the comprehension of what sort of justice the environment receives and what sort of model is presented to the reader who is learning good citizenship. To unpack frame packages (Stibbe, 2015; Venkataraman, 2020), finding collections of beliefs, procedures and practices patterned on grammatical, rhetorical and stylistic devices was the task. Framing occurs when a frame is used to package a particular part of life through 'trigger words' that like a metaphor impose one pattern onto another to describe that area anew. Each article in the dataset was annotated for trigger words to detail rhetorical, syntactic and thematic structures

(Pan & Kosicki, 1993). These 'interpretative' packages or frames disclosed how *Taifa Leo* told at a tale of climate change in Kenya.

Results/findings

As Table 8.1 indicates, material clauses constitute 49 per cent of the processes. Following that, mental processes consume 18 per cent of the clauses, and relational clauses constitute up to 16 per cent. The first three columns show processes that together encompass 83 per cent of all the clauses, while verbal processes (13 per cent) and existential processes (4 per cent) comprise only 17 per cent of the whole of the texts. Given that the number and percentage of the verbal and existential processes are few and far between, they reflect less significance to the newspaper reports and what we can say about the two. As a consequence, our discussion concentrated on analysing the human material, mental and relational processes to uncover the experiential meaning of the articles.

Table 8.2 reveals that none of the ten newspaper articles challenge the canonical paradigm that human beings dominate nature. When natural participants occur, they are predominantly Tokens (51 per cent) or Affecteds (41 per cent). As Actor Transitive, they threaten human life: *mbu wanaosababisha malaria* 'malaria-causing mosquitoes' (Ongaji, 2022). As Actor Intransitive, they serve as a resource: *hadi miwa iliyo shambani ikomae* 'until the sugarcane in the field ripens' (Makokha, 2023). Though the report depicts *miwa* in action, its action is of ripening to produce sugar for human consumption or perhaps to produce fuel, which marks it clearly as a resource whether for energy or food. As a consequence, the analysis shows that the majority of clauses depict nature as a natural participant serving the needs of humanity by producing, providing and supplying produce, goods and services for the sustenance and maintenance of humankind.

Scarcity of depictions of non-human entities in the data formed a pattern of erasure and encouraged us to turn to framing as the tool to analyse the ecosophy (Stibbe, 2015) or ecological philosophy being modelled as a basis for forming good citizenship. Table 8.3 provides an overview of the issues, exemplars and devices that trigger and categorize the framing. CDA allowed us to look beyond texts and to consider the historical and sociocultural contexts. In the analysis of language and rhetoric, we examined the formulations advanced by social actors and in the discourse of news reporters. The discursive strategies are forms of discourse manipulation of reality in order to achieve a certain effect or goal.

Table 8.2 Natural Participants in Clauses

	Affected	Actor Transitive	Token	Actor Intransitive
A1	Misitu (84)	0	Miti (84), za kilimo (84)	0
A2	0	0	0	0
A3	Mito (38), mimea (61), mifugo (106, 107)	Mbu (67)	Wanyama (77, 83), wadudu (87), mwanabiolojia (87), viumbe hai (88, 90)	0
A4	Konokono (2, 13)	0	Mpunga (4), wadudu (6)	0
A5	Viumbe (6)	0	0	0
A6	0	0	Misitu (8), msitu (42)	0
A7	0	0	0	0
A8	Miti (19, 22, 23, 24, 25, 28)	0	0	0
A9	Miti (8)	0	0	0
A10	Miwa (17)	0	Miwa (3, 6, 7, 14, 21, 54), majanichai (39), kahawa (39)	Miwa (18, 19)
Totals	16/39 (41%)	1/39 (2.5%)	20/39 (51%)	2/39 (5%)
Percentage	16/498 (3.2%)	1/498 (0.2%)	20/498 (4%)	2/498 (0.4%)

Table 8.3 Frames of Ecological Contention

Issues	Exemplars	Devices
Colonization	*mataifa yanayostawi* 'developing nations'	Development
Westernization	*Suluhisho la Fedha* 'finance solutions'	Economic problem
Tradition	*machozi ya furaha* 'tears of joy'	Nostalgic resource
Responsibility	*madai* 'allegations'; *lawama* 'blame'	Modern predicament

The notion of discursive strategies helped us perceive the link between the context and issue representation. Two consequences emerged from the issues classified. First were the pervasive patterns of an anthropocentric view of the environment, and second was a cultural link between past experiences and contemporary perceptions of reality. We explored the paper's modelling of the issue of colonization through a lens of development towards Westernization as an economic problem that led to a nostalgic view of traditional engagement with the environment as a resource that ultimately consciously raises the perspective that the modern predicament of climate change requires a local conception of responsibility to safeguard ecosystems.

Discussion

Our discussion does not explore all the themes or frames modelled in *Taifa Leo*; rather the endeavour is to seek to illustrate how the frames contribute to a model of ecological identity citizen readers may emulate. Our analysis reinforces Stibbe's (2015) framework for analysing the language of advertisements promoting multinational companies. The aim of the chosen conceptual framework is to disclose and, if necessary, defy, the narratives surrounding the audience. These narratives are what Stibbe (2015) called 'stories we live by' because they are the cognitive models that become embedded in the minds of readers through exposure. Newspaper stories only disclose hidden narratives after meticulous examination because only then do clues, hints and triggers disclose the cognitive structures found in the stories by which we live our lives. As to how stories wax or wane destructive or beneficial to ecology, their texts present themselves in one of eight forms: conviction, erasure, evaluation, metaphor, ideology, salience, framing and identity. The present investigation concentrates on identity but considers the construction of framing as the key to modelling for citizenship in the newspaper.

Decolonization

Frames prove to be essential to the discipline of ecolinguistics because the concept of framing frequently provides a path to manifesting 'social change' (Stibbe, 2015: 40). If a frame is understood as a package of knowledge and frames are collective cognitive knowledge, then a community has social knowledge unique to their culture. In fact, Gavriely-Nuri (2012: 78) defines culture as a database that packs shared social beliefs, ethos and values as an 'open-ended repertoire' accessible to a community.

For instance, navigating the Kenyan cultural database around the concept of 'development' leads, among other issues, to the struggle for independence.

Examples walk through the cultural code packaging: Part (a) is original Swahili wording; part (b) is a liberal English translation.

Original Text	**English Version**
(1a) *Mataifa yanayostawi yamekuwa yakishinikiza mataifa matajiri kuhakikisha kwamba ufadhili wa hasara inayotokana na majanga yanayosababishwa na mabadiliko ya tabianchi, unafikia angaa dola 100 bilioni kila mwaka . . .* (Ongaji, 2022)	(1b) Developing countries have been pushing rich countries to ensure that the financing of losses resulting from disasters caused by climate change reaches 100 billion dollars annually . . .

The phrase *mataifa yanyostawi* 'developing nations' triggers the frame of development that entails a theme of decolonization. Sachs (1999: 29) points out that 'development' itself has had the insidious effect of dissolving cultures that were based on something other than the accumulation of riches. The framing is frequent in political discourse (Stibbe, 2015: 55), and it is significant here because of the impact on local ecosystems and social justice. Decolonization encompasses the removal of Western biases and worldview that became ingrained in 'developing' countries and institutions during colonialism (du Toit, 2023; Gavriely-Nuri, 2012). In addition, the frame triggers the symbolism of resistance in the national anthem (Mwinlaaru & Nartey, 2022) and the national flag. The Kenyan discourse on the struggle for independence includes the ethos of the *yakishinikiza mataifa matajiri* 'pushing of rich nations' off native lands. It is the myth of a historical freedom-fighting narrative that is symbolized in the country's flag. The Kenyan national flag consists of three equal-sized horizontal bands of colour with green at the bottom, red in the middle and black on the top.

A Maasai traditional shield with two crossed spears supporting it from behind stands in the centre of the flag, transecting all three bands of colour. In his autobiography, former freedom-fighter *Karari wa Njama* elaborates on how Kenya's future first president described the flag's symbolism:

> What he [Jomo Kenyatta] said must mean that our fertile lands (green) could only be regained by the blood (red) of the African (black). That was it! The black was separated from the green by the red: The African could only get to his land through blood.
>
> (Barnett & Njama, 2021)

The symbols of fighting (the shield and spears) transect the bands and bridge the people through their blood to nature. The connections between the strata of grammar, discourse and social context similarly relate the symbolism of the flag to the codes of the culture through what systemic-functional grammar calls realization (Martin & Rose, 2003). Freedom and development in Kenya are linked in part to these cultural codes.

Development

A daunting aspect of the framing of development also consists of a predicament as underscored in Example 2:

Original Text	**English Version**
(2a) . . . *licha ya kuwakilisha chini ya asilimia 4 pekee ya uzalishaji wa gesi hizi hatari, Afrika inaendelea kuathirika pakubwa na janga hili huku wataalamu wakitabiri ongezeko la halijoto barani ikilinganishwa na sehemu nyingine zozote ulimwenguni.* (Ongaji, 2022)	(2b) . . . despite representing only less than 4 per cent of the production of these dangerous gases, Africa continues to be greatly affected by this epidemic as experts predict an increase in temperature in the continent compared to any other part of the world.

Though developing countries like those in Africa produce less than 4 per cent of global warming gases, they are paying higher costs than developed countries. More than 3.5 million people in over twenty-three counties in Kenya alone are facing hunger (Ongaji, 2022). Stibbe (2015) warns that framing this approach as a problem summons up the idea that there is a solution. The solution suggested is that developing countries need to 'develop' more, but as Example 2 suggests, the dangerous global warming gases are being produced by *mataifa yaliyostawi* 'developed nations'. Consider, for instance, the second half of the theme for the first Africa Climate Summit (ACS) in Example 3:

Original Text	**English Version**
(3a) . . . *na Suluhisho la Fedha za Mabadiliko ya Tabianchi Barani Afrika na Dunia* (Waweru, 2023)	(3b) . . . and Climate Change Finance Solutions in Africa and the World

Though it is arguably admirable to want solutions to global warming, framing climate change as a problem is a 'false hope' (Cachelin et al., 2010) because the underdeveloped countries are viewed as the ones who must change when it is already acknowledged that the developed countries, as in Example 4, are the ones who have created the circumstances *kupitia ustawi wao* 'through their prosperity'.

Original Text	**English Version**
(4a) *Utafiti wa kisayansi umeonyesha kwamba mataifa yaliyostawi ndio yamechangia pakubwa uzalishaji wa gesi hizi ikiwa ni pamoja na dioksidi ya kaboni kupitia ustawi wao*. (Ongaji, 2022)	(4b) Scientific research has shown that developed nations have contributed greatly to the production of these gases including carbon dioxide through their prosperity.

This very scenario, like the *Kuinua Ukuaji Kijani* 'Raising Green Growth' appeal, sets up profit-making or economic growth as the preferred advice. Such a situation could lead to a consumer mentality or what Alcott (2005) labels as Jevon's paradox. This situation is a rebound effect in that if consumers or developing countries raise more and more *fedha* or 'finance' in chasing development and Westernization, there will not be *chini ya asilimia 4 pekee ya uzalishaji wa gesi hizi hatari* 'less than 4 percent of the production of these dangerous gases' (Ongaji, 2022). There will be more! The danger of the plunder of old imperialism (i.e, exploitation for foreign profit) revisits as the allure and delusion of 'equitable development' (Sachs 2010: 28). Resources become depleted, ecosystems damaged and the legacy for posterity in ruins (Stibbe, 2015: 53).

Nostalgia

A legacy of ruins for posterity is enough to make anyone, even the Kenyan everyman, frame the situation as an apocalyptical tragedy or 'a linear temporality emphasizing a catastrophic endpoint . . . outside the purview of human agency' (Foust & O'Shannon, 2009). Throughout the 1930s' novel *Green Hills of Africa*, Hemingway again and again describes Africa as a continent already falling into the same environmental demise as that of the Global North (Tyler, 2017: 39). The educated and uneducated alike may be lulled into revelling in nostalgia, as Example 5 illustrates.

Original Text	**English Version**
(5a) *Kwa mfano, nyanyangu angeeleza kiurahisi kwamba msimu wa upanzi ulikuwa kati ya mwezi Machi na Mei. Kwa kauli yake, ni wakati huo ambapo 'miungu walitiririsha machozi ya furaha' kwa jamii. Urejeleo wake unatokana na uzamani ambao mababu zetu walizingatia katika utabiri wa mikondo ya nyakati za hali ya hewa. Hawakukosea. Miungu walikuwa na furaha. 'Wangetiririsha machozi' yao wakati ambao ulifaa, bila kuchelewa. Lakini sasa? Mambo ni tofauti sana.* (Kamau, 2023)	(5b) For example, my grandmother would simply explain that the planting season was between March and May. According to her, it was at that time that 'the gods shed tears of joy' for the community. Its reference comes from the ancients that our ancestors considered in predicting weather patterns. They weren't wrong. The gods were happy. They would 'shed their tears' at the right time, without delay. But now? Things are very different.

Framing of the elements of climate change in contrast to the way the climate and the environment were in the past still presents a nostalgized situation or *wakati huo* 'that time' in a frame of nature as a resource. The source frame is clearly triggered by the phrasing of *machozi ya furaha* 'tears of joy' as shed in exchange for something, an implicit *kafara* 'sacrifice'. The *machozi* 'tears' are the rain for the soil for the crops to grow. As a resource, rain becomes an object owned by an owner, the cultivator, who has the right to do as she/he pleases. The implicit ecological issue here is that the framing labels the resource as a thing to be exploited. But now *miungu* 'the gods' are not holding up their end of the bargain, so the cultivator is being denied her/his right, being punished. Granted, if the cultivator is a subsistence farmer, the issue is a matter of life or death. Hence, powerlessness and desperation lead to an apocalyptic frame. Mapped onto climate change, this frame causes resignation and perhaps the blame game, neither of which assist in the alleviation of the impact of climate change nor create resilience in a community to adapt to an inevitable reality.

Reality

Change has come. That is the reality. It is a modern predicament. Framing climate change as a predicament allows for two manageable elements: the predicament and the response. Though the predicament cannot be solved, it can be dealt with. Knowing that there is a way forward changes the mindset of the cultivator and anyone facing the predicament anew.

A former predicament was that Kenya, along with many other African and other nations of the Global South, was dominated by colonial powers. A foreign military and religion were leveraged to suppress and denigrate indigenous cultural and religious traditions precipitating social decline and environmental devastation. That predicament was responded to and mitigated against through a reassertion of cultural pride and resilience via a struggle for independent governance.

Original Text	English Version
(6a) *Kinaya cha mwanadamu wa sasa na yule wa kizamani ni kuwa, wale wa 'zamani' waliyajali mazingira. Hata ikiwa walikata miti, waliikata na kupanda mingine.* (Kamau, 2023)	(6b) The irony of the modern man and the ancient one is that those 'of the past' cared about the environment. Even if they cut down trees, they cut them down and planted others.

Example 6 shows how contemplation of the way things are compared to the way things were could lead to an ascription of blame. The implication of *yule wa kizamani* 'the ancient one' in contrast to *mwanadamu wa sasa* 'the modern man' can be *kinaya* 'irony' only if the former is somehow better than the latter. Implicit in being modern is the connotation of being better. Since modern life is ironically far worse than past life, a reason is sought for explanation. There is no doubt that a revival of respect for and practice of indigenous traditions can promote the conservation of nature and the environment. However, a challenge arises if an attitude of absolutism comes to view all tradition as good, and everything modern or Western as bad.

Original Text	English Version
(7a) *Zaidi ya hayo, kuna miti ambayo haingekatwa, kwani ilichukuliwa kwa uzito mkubwa kama chemichemi ya baraka kwa jamii. Hilo lilihakikisha kuwa, hata ikiwa kulikuwa na ukataji miti, chemichemi muhimu za maji hazikukauka. Kinyume cha hayo, msukumo wa mwanadamu wa sasa ni tamaa tupu.* (Kamau, 2023)	(7b) Furthermore, there are trees that would not be cut down, as they were taken very seriously as a source of blessing for the community. This ensured that, even if there was deforestation, important aquifers did not dry up. In contrast, the current human impulse is pure desire.

If the *msukumo wa mwanadamu wa sasa* 'the modern human impulse' of Example 7 is *tamaa tupu* 'pure lust', rejection of everything modern or Western is an easy resolution. Such an approach rejects modern ways for their treatment of the ecosystems as a commodity and a means to profit, a desacralization of nature. Revival of native traditions treats the environment as sacred and deserving of respect, what Taylor (2013: 190) calls 'dark green' spirituality. This spirituality may or may not include a belief in the divine and the supernatural. It is 'green' for tending to view nature and all life in our biosphere as sacred. It skews 'dark' because it frequently comes with an apocalyptic expectation that adherents have a 'religious duty' to resist or usher in the coming climate change or to defend the environment as sacred (Taylor, 2010: 84). Put differently, holy ecological war is implicit. From this brief review, we can see that the objectives of dark green religion not only restrict what is accepted as native but further potentially promote biases by turning a blind eye to actions that could mitigate climate change impacts, avoid adaptations to change or cause potentially resilient communities to lose hope. The framing risks moving Western thought, including science, away from potential partnership and support, and into the role of oppression, and it depicts aid as yet another form of colonialism.

In Kenya, climate change has exacerbated drought and caused conflict among communities and social unrest over resources (Njiru, 2012). It has even been asserted that the 2007 post-election violence was about struggles to retain or gain land (Taylor, 2013: 187). The risk of conflict is so high that even the current president recently felt the need to address land grabbing, as seen in Example 8.

Original Text	**English Version**
(8a) *Hata hivyo kuhusu madai kuwa kuna baadhi ya misitu kama Mau kuuziwa wawekezaji wa hewa ukaa, Rais Ruto amekanusha kuwa hakuna hata inchi moja au kipande cha msitu zimeuziwa wawekezaji nchini Kenya.* (Walter, 2023)	(8b) However, regarding the claim that there are some forests like Mau that have been sold to carbon investors, President Ruto has denied that not even an inch or piece of forest has been sold to investors in Kenya.

The phrasing *hakuna hata inchi moja au kipande cha msitu* 'not even an inch or piece of forest' makes clear that rumours about land being sold to foreigners were credible and worrisome enough that the president had to deny *madai* 'allegations' to avoid a potential eruption of violence. But this threat is only one symptom. Drought is desiccating rivers, killing domesticated animals and starving predators in the wild, who then turn to prey on domestic animals. The climate change consequence is 'a vicious feedback loop' of environmental degradation that is starving people as well (Taylor, 2013: 187). Yet tillable and grazable land shrinks as the population continues to grow. Tradition can encourage even Kenya's educated elite to minimize the reality of the threat to future peace and social stability from a continuing rapid increase in population growth. This predicament of tradition versus Western cultures (Taylor, 2013: 197) ultimately hinders action because African environmentalists may be dismissed as anti-scientific and naïve, while Western-educated African scientists may be condemned for being sell-outs to colonial masters.

Example 9 demonstrates that though the Global North bears responsibility for the global warming affecting the world, the Global South is not completely innocent. The governments of Africa are not without *lawama* 'blame' themselves.

Original Text	**English Version**
(9a) *Lakini huku mataifa ya Afrika yakiendelea kushinikiza nchi zilizostawi kuwajibika katika masuala haya ya mazingira na mabadiliko ya tabianchi, pia serikali za hapa barani haziepuki lawama.* (Ongaji, 2022)	(9b) But while African nations continue to press developed countries to take responsibility in these environmental issues and climate change, the governments here on the continent do not escape blame.

The climate change predicament is constructed through a listing of *shughuli za uharibifu wa mazingira nyumbani pia zimechangia* 'environmentally destructive activities at home have also contributed significantly' (Ongaji, 2022). What drives climate change at home is the framing of the environment as a resource, something to be exploited by human beings.

Original Text	**English Version**
(10a) *Wataalamu wanasema kwamba kwa kiwango fulani, shughuli za uharibifu wa mazingira nyumbani pia zimechangia pakubwa. Kwa mfano, ukataji haramu wa miti, mbinu duni za kilimo, na uvamizi wa ardhi iliyotengewa misitu kunachangia hali inayoshuhudiwa.*	(10b) Experts say that to some extent, environmentally destructive activities at home have also contributed significantly. For example, illegal logging, poor agricultural practices, and encroachment of land reserved for forests contribute to the situation being witnessed.

This *kuwajibika* 'being responsible' modern predicament is an issue-frame proposing a path to mitigation. Taylor (2013: 196) offers a way around the 'epistemological myopia' of traditional culture versus Western culture. An essential attribute of an ecologically beneficial identity is valuing all living entities in our biosphere as interdependent ecologically. We can avoid a 'genetic fallacy' if we see the value of a mitigation strategy as not dependent on its origin. Globalization spreads, the social world merges and tradition blurs into the Western. Our quest to exist must follow a pragmatic path to hybridize human knowledge, human practice and human experience so that we meet the gold standard of nature – that is, we survive by protecting ecosystems.

Conclusion

In sum, this study investigated the potential role of *Taifa Leo*'s reporting on *mabadiliko ya tabianchi* 'climate change' in modelling a beneficial ecological identity for good citizenship in Kenya. To our knowledge, it is the first study to demonstrate the ecosophy of Kenyan Swahili news reports to influence the burgeoning ecological identities of readers. In the sampling of ten articles, the results indicate that the language of reportage influences in a mostly destructive or ecologically adverse manner. An SFG examination of the natural participants and transitivity of clausal processes showed that 49 per cent of clauses are Material process verbs that describe actions in which a human Actor manipulates an affected non-human object or recipient. On the one hand, if natural Transitive Actors occur, they endanger human beings. On the other hand, if natural Actors are Intransitive, they function as resources for human society. This anthropocentric perspective is in line with previous studies analysing the biases against natural participants in the media (Stibbe, 2015; Atieno & Njoroge, 2016; Poore & Nemecek, 2018; Venkataraman, 2018). The results are less clear and less destructive but perhaps more ecologically beneficial with respect to the influence of the gazette's projecting Kenya's history of *mabadiliko ya tabianchi* 'climate change' as an issue frame to describe a cultural code of chronology as nation building. The fact that the issues of decolonization and development function together to guide citizen readers through a tailored narrative that exploits both issues of nostalgia and patriotic responsibility to encourage behaviours that support and assist ecosystems is profound. Our study indicates that these newspaper articles positioned readers as wanting their country's economy to be environmentally ethical, sustainable and non-human-species-friendly for a beneficial impact on the world. As modern Kenyan readers see modelled intrinsic values that centre on the global ecology, these newspaper articles are somehow forging a new Kenyan ecological identity as ethical participants rather than emulating the foolhardy following of the selfish Western development model of yore.

There may be some limitations to this study with regard to the collection of data and interpretation of the results. A first limitation could be the omission of further news articles and/or articles from more than one Swahili newspaper. For example, additional data could have come from *Mwananchi*, Tanzania's leading Swahili paper, or even from Kenya's English *Daily Nation* and *The Standard*. Another potential shortcoming in the study is a bias in methodological approach. We used cultural codes and issue framing to chronicle a history of Kenyan events to construct an overarching shift in ecological views from anthropocentric and resource-centred to biospheric and globally focused, so perhaps the relationship link between cultural codes, the independence struggle and apocalyptic expectation is a bit inflated. The confidence in our findings could be strengthened with data from ecological identity surveys from actual students and teachers engaged in the Newspapers in Education Programme (NiE). It would then be possible to examine how *Taifa Leo* and schoolteachers model identities in students' minds that encourage behaviours that protect ecosystems. Of course, these shortcomings could inspire future researchers to better define their own research agenda.

References

Alcott, B. (2005). Jevons' paradox. *Ecological Economics, 54*(1), 9–21.

Atieno, L., & Njoroge, J. M. (2016). A review of the "Tourism-Environmental Conflict Signifier" logic as constructed by the Kenyan press in coverage of environmental conflicts. *Language & Ecology*. Retrieved from: www.ecoling.net

Barnett, D., & Njama, K. (2021). *Mau Mau from within: The story of the Kenya land and freedom Army from within* (New and Expanded). Daraja Press.

Bylanes, K. (2021, November 10). *Nairobi news: Taifa Leo set for relaunch* [Nairobi news: Nation today set for relaunch]. Retrieved from https://nairobinews.nation.africa/taifa-leo-set-for-relaunch/

Cachelin, A., Norvell, R., & Darling, A., (2010). Language fouls in teaching ecology: Why traditional metaphors undermine conservation literacy. *Conservation Biology, 24*(3), 669–674.

Dagnet, Y., Waskow, D., Bergen, M., Levin, K., Ringuet, L. N., Thwaites, J., Holt, M., Cogswell, N., Elliott, C., Bouyé, M., & Worker, J. (2019, December 2023). COP25:

What we needed, what we got, what's next. *World Resources Institute*. Retrieved from https://www.wri.org/insights/cop25-what-we-needed-what-we-got-whats-next

Du Toit, L. J. (2023). Investigation into the development of a methodology for the study of environmental discourses. *HTS Teologiese Studies/Theological Studies*, *79*(3), a8105. https://doi.org/ 10.4102/hts.v79i3.8105

Foust, C., & O'Shannon, M. W. (2009). Revealing and reframing apocalyptic tragedy in global warming discourse. *Environmental Communication: A Journal of Nature and Culture*, *3*(2), 151–167.

Gardner, J., & Sullivan, B. (1999). The national newspaper as a tool educational empowerment: Origins and rationale. In S. R. Knowlton & B. O. Barefoot (Eds.), *Using national newspapers in the college classroom: Resources to improve teaching and learning* (pp. 11–17). National Resource Center for the First-Year Experience & Studentsin Transition. New York Times Co., NY. (ED433745). ERIC. https://files.eric.ed.gov/fulltext/ED433745.pdf

Gathara, P. (2018, August 02). Black, Red, and Green: The story behind the Kenyan flag. *The Elephant –African Analysis, Opinion, and Investigation*. Retrieved from https://www.theelephant.info/features/2018/08/02/black-red-and-green-the-story-behind-the-kenyan-flag/

Gavriely-Nuri, D. (2012). Cultural approach to CDA. *Critical Discourse Studies*, *9*(1), 77–85.

Halliday, M., & Matthiessen, C. (2014). *Halliday's introduction to functional grammar* (4th ed.). Routledge.

Kamau, W. (2023, April 08). *Taifa Leo: Tamaa Ya Mwanadamu Ni Chanzo Cha Uharibifu Wa Mazingira* [Nation today: Human desire is the source of environmental destruction]. Retrieved from: https://taifaleo.nation.co.ke/maoni/wanderi-kamau-tamaa-ya-mwanadamu-ni-chanzo-chauharibifu-wa-mazingira

Kitonga, N. N., Owala, S. O., & Ombati, J. M. (2023). Choice of language in the reporting of environmental news in Taifa Leo newspaper in Kenya. *International Journal of Research-GRANTHAALAYAH*, *11*(10), 100–112. https://doi.org/10.29121/granthaalayah.v11.i10.2023.5343

Lakoff, G. (2010). Why it matters how we frame the environment. *Environmental Communication: A Journal of Nature and Culture*, *4*(1), 70–81.

Larson, B. (2011). *Metaphors for environmental sustainability: Redefining our relationship with nature*. Yale University Press.

Londoño, E. (2019, October 21). What you need to know about the unrest in chile. *New York Times*. Retrieved from https://web.archive.org/web/20191101174629/https://www.nytimes.com/2019/1

Makokha, S. (2023, August 01). *Taifa Leo: Gavana Sasa Alaumu Kampuni Ya Uganda Kwa Uhaba Wa Miwa* [Nation today: Governor now blames Ugandan company for sugarcane shortage]. Retrieved from: https://taifaleo.nation.co.ke/habari/habari-za-kitaifa/gavana-sasa-alaumu-kampuni-ya-uganda-kwa-uhaba-wa-miwa

Martin, J., & Rose, D. (2003). *Working with discourse: Meaning beyond the clause.* Bloomsbury.
Mautner, G. (2008). Analyzing newspapers, magazines and other print media. In R. Wodak & M. Krzyzanowski (Eds.), *Qualitative discourse analysis in the social sciences* (pp. 30–51). Palgrave Macmillan.
Mukhwana A. S., & Mukhwana C. S. (2017). The role of Kenya's *Taifa Leo* newspaper in enhancing learners' abilities in Kiswahili language proficiency. *International Journal of Humanities Social Sciences and Education* (IJHSSE), *4*(12), 53–61.
Mwinlaaru, I., & Nartey, M. (2022). 'Free men we stand under the flag of our land': A transitivity analysis of African anthems as discourses of resistance against colonialism. *Critical Discourse Studies, 19*(5), 556–572.
Njiru, B. (2012). Climate change, resource competition, and conflict amongst pastoral communities in Kenya. In Jürgen Scheffran, Michael Brzoska, Hans Günter Brauch, Peter Michael Link, Janpeter Schilling (Eds.), *Climate change, human security, and violent conflict challenges for societal stability* (pp. 513–527). Hexagon Series on Human and Environmental Security and Peace, vol. 8. Springer.
Ongaji, P. (2022, November 08). *Taifa Leo: Mabadiliko Ya Tabianchi: Afrika Pia Sharti Iwajibike* [Nation today: Climate change: Africa must be responsible]. Retrieved from: https://taifaleo.nation.co.ke/makala/mabadiliko-ya-tabianchi-afrika-pia-sharti-iwajibike
Painter, J. (2013). *Climate change in the media: Reporting risk and uncertainty.* Reuters Institute for the Study of Journalism.
Pan, Z., & Kosicki, G. (1993). Framing analysis: An approach to news discourse. *Political Communication, 10*(1), 55–75.
Poore, J., & Nemecek, T. (2018). Reducing food's environmental impacts through producers and consumers. *Science, 360*(6392), 987–992.
Rowlatt, Justin (2023, November 27). UAE planned to use COP28 climate talks to make oil deals. *BBC News.* https://www.bbc.co.uk/news/science-environment-67508331
Sachs, W. (1999). *Planet dialectics: Explorations in environment and development.* Zed Books.
Sachs, W. (ed.). (2010). *The development dictionary: A guide to knowledge as power* (2nd ed.). Zed Books.
Sengupta, S. (2019, October 21). Spain agrees to host key climate talks after Chile pulls out. *New York Times.* Retrieved from: https://web.archive.org/web/20191031220450/https://www.nytimes.com/2019/10/31/climate/spain-chile-cop25-climate-meeting.html
Shaw, C., & Nerlich, B. (2014). Metaphor as a mechanism of global climate change governance: A study of international policies, 1992–2012. *Ecological Economics, 109*, 34–40.
Stibbe, Arran. (2015). *Ecolinguistics: Language, ecology and the stories we live by.* Routledge.

Taylor, B. (2010). *Dark green religion: Nature, spirituality, and the planetary future*. University of California Press.

Taylor, B. (2013). Kenya's Green Belt Movement: Contributions, conflict, contradictions, and complications in a prominent Environmental Non-Governmental Organization (ENGO). In T. Lars, B. Witoszek, & B. Taylor (Eds.), *Civil society in the age of monitory democracy* (pp. 180–207). Berghahn Books.

Tyler, L. (2017). "How the weather was": Anthropogenic climate change and environmental damage in Hemingway's *green hills of Africa. The Hemingway Review*, *37*(1), 36–54.

Van Dijk, T. (ed.). (2011). *Discourse studies: a multidisciplinary introduction* (2nd ed.). Sage.

Venkataraman, N. (2018). What's not in a frame? Analysis of media representations of the environmental refugee. In M. Schröter & C. Taylor (Eds.), *Exploring silence and absence in discourse: Empirical approaches* (pp. 241–279). Springer

Venkataraman, N. (2020). The absent victims: An ecolinguistic study of environmental refugees in The New York Times. *JAM It! (Journal of American Studies in Italy)*. May (3), 109–141.

Walter, J. (2023, December 03). *Taifa Leo:Serikali Kupendekeza Sheria Ya Kudhibiti Soko La Hewa Ukaa Nchini Kenya* [Nation today: The government proposes a law to regulate the residential air market in Kenya]. Retrieved from: https://taifaleo.nation.co.ke/habari/habari-za-kitaifa/serikali-kupendekeza-sheria-ya-kudhibiti-soko-la-hewa-ukaa-nchini-kenya

Waweru, S. (2023, September 03). *Taifa Leo: Dereva: Rais Ruto Aendesha Gari Linalotumia Nguvu Za Umeme Kuingia KICC, Nairobi* [Nation today: Driver: President Ruto drives an electric vehicle into KICC, Nairobi]. Retrieved from: https://taifaleo.nation.co.ke/habari/habari-za-kitaifa/dereva-rais-ruto-aendesha-gari-linalotumia-nguvu-za-umeme-kuingia-kicc-nairobi

9

A biographical approach to decolonial ecolinguistics

Learning from the margins about endangered ecologies in Senegal

Monika Christine Rohmer and Mouhamed Abdallah Ly

Abstract:

With our contribution, we want to advocate for an inclusion of marginal biographies and voices in ecolinguistics. On the one hand, we show how life stories on the margins can contribute to decolonising environmental knowledge. On the other hand, we shed light on the entanglements of marginal lives in contemporary modernity. Our chapter departs from autobiographical accounts of Ibrahima Diop, also known as Iran, a Senegalese man who for a long time had the reputation of being a criminal or 'bandit'. The reputation as a criminal was bestowed on Diop by those seeing him for many years in the company of a pack of dogs, heading through the peri-urban forests of Senegal's capital Dakar. These forests and non-urbanized lands within the city constitute a refuge for many marginalized people, including delinquents. It was only recently, thanks to social media, that the general public was able to discover a relatively unknown facet of Diop. The life of this individual in the peri-urban forests consists of growing vegetables for his own needs, taming anacondas, and experimenting with plant-based healing methods based on indigenous ecological knowledge. Our contribution is part of a larger effort of putting into question globalized discourses through the analysis of texts in African languages in and beyond Senegal. In this contribution, we are looking at the 'marginalized' biography of Ibrahima Diop, narrated in a 'marginalized' language, namely Wolof, and we reflect on the theoretical implications of this narration for decolonial ecolinguistics.

Key words: indigenous voices, marginality, Southern epistemologies, Senegal, Wolof language.

Introduction

Our contribution is in line with recent academic endeavours of learning from biographical accounts of individuals, especially those situated on the margins of society (Daha, 2002; Faye & Thioub, 2003). Thereby, we intend to give a voice to people often disregarded in traditional academic approaches or in surveys designed for public policy purposes. With our contribution, we want to advocate for the inclusion of marginal biographies and voices in ecolinguistics. On the one hand, we show how life stories on the margins can contribute to decolonizing environmental knowledge. On the other hand, we shed light on the entanglements of marginal lives in contemporary modernity. Thus, there are limitations to interpreting marginal African biographies from a decolonial perspective.

Our chapter departs from autobiographical accounts of Ibrahima Diop, also known as Iran, a Senegalese man who for a long time had the reputation of being a criminal or 'bandit'. The reputation as a criminal was bestowed on Diop by those seeing him for many years in the company of a pack of dogs, heading through the peri-urban forests of Senegal's capital Dakar. These forests and non-urbanized lands within the city constitute a refuge for many marginalized people, including delinquents. It was only recently, thanks to social media, that the general public was able to discover a relatively unknown facet of Diop. The life of this individual in the peri-urban forests consists of growing vegetables for his own needs, taming anacondas and experimenting with plant-based healing methods based on indigenous ecological knowledge.

In this chapter we propose a reading of Diop's narration about his way of life as a form of decolonial resistance by applying the theoretical reflections of Arturo Escobar (2016, 2018) and Malcom Ferdinand (2021, 2019). We suggest that the unique relation of Diop with his convironment[1] invites ecological reflection and rethinking. His way of life and self-narration counter current trends in Senegal and other countries in the Global South to follow a developmentalist agenda and ideology at the expense of indigenous knowledge. Our contribution is part of a larger effort of putting into question globalized discourses through the analysis of texts in African languages in and beyond Senegal (Ly & Rohmer,

2024). Hence, in this contribution, we are thinking at the edge of linguistics and language science. We are looking at the 'marginalized' biography of Ibrahima Diop, narrated in a 'marginalized' language, namely Wolof, and we reflect on the theoretical implications of this narration for decolonial ecolinguistics.

Theoretical approach

Escobar's relational ontology and Ferdinand's decolonial ecology

We fully agree with the aims spelt out by the International Ecolinguistics Association of addressing contemporary ecological challenges in linguistics and developing theories that acknowledge the embeddedness of humans in the web of life. However, to do ecolinguistics in contemporary West Africa also necessitates considering the post-colonial context. Since organizing a workshop in February 2023 on ecolinguistics in Senegal, we have been reflecting on the theoretical implications of doing ecolinguistics in a post-colonial African context. While we adhere to the primacy of ecolinguistics that a thorough understanding of language also requires looking at the situatedness of speakers in their environment, in doing linguistics in Senegal, another dimension is necessary.

In Senegal, the interrelations of language and ecology are marked by colonial and post-colonial encounters. Most obviously, this can be seen in the linguistic landscape of Senegal. Wolof is the language of everyday communication in Senegal. The language, belonging to the Senegambian branch of the Niger-Congo language family, is spoken by roughly 90 per cent of the population (Diouf et al., 2017; McLaughlin, 2017). However, the only official language of Senegal since colonization is French. French is only spoken by a minority of the population, with figures ranging between 20 and 25 per cent of the population (Cissé, 2005; Diouf et al., 2017; McLaughlin, 2008). This shows that Wolof is not a *marginal* language in Senegal but *marginalized*. While Wolof is spoken and used widely,[2] the language of formal education, state administration and institutionalized lawsuits remains French.

The contemporary linguistic landscape shows that it is necessary to take into account colonial history in the analysis of contemporary Senegal. Approaches to decolonization, very schematically, depart from the assumption that the contemporary global system privileges a centre that corresponds to the former colonizing nations. Through colonization, an economic system of international capitalist division of labour was established, but worldviews were also implemented that have upheld unjust economic distribution until today. Therefore, domination continues to characterize relations between former metropolitan countries and former colonies. Decolonial approaches criticize a continuous interiorization and systematic exploitation of racialized people by a hegemonic system based on the coloniality of being, power and knowledge. In this chapter, our conceptualization of decolonization follows closely the work of Arturo Escobar and Malcom Ferdinand.

In reading and analysing Ibrahima Diop's narrative, we are guided by Arturo Escobar's (2016, 2018) reflections on the possibility of a *plurivalist world* and his concept of *relational ontologies*. Extending Boaventura de Sousa Santos's work on epistemologies of the Global South, Escobar's reflections are inspired by practices of resistance to today's dominant framework, or, in the words of Escobar, a 'One-World' world. The One-World ideology and project for Escobar is defined and working through two ontological divides: (1) the 'colonial divide', separating those inside and at the margins of the One-World world and (2) the 'nature/culture divide' separating humans from nature (Escobar, 2016, p. 21). Therefore, forms of territorial resistance against One-World worlding are to be 'understood as ontological struggles and as struggles over a world where many words fit' (Escobar, 2016, p. 13). For the vision of a world that accommodates many worlds, he uses the term 'pluriverse'. To establish this pluriverse, Escobar (2016, p. 13) requires researchers to learn as much 'from the experience, knowledge, and struggles of subaltern groups as from the academy'.

The two ontological divides outlined by Escobar lead to what Malcom Ferdinand (2021, 2019) describes as a 'double fracture'. This concept highlights the interconnected crises stemming from colonial legacies and ecological degradation. The double fracture reflects the complex interplay between social injustices and environmental challenges, illustrating how these divides not only shape our understanding of the world but also exacerbate existing inequalities and ecological vulnerabilities. Ferdinand criticizes contemporary environmental and ecological practices that neither acknowledge nor resist the colonial past, while anticolonial struggles 'forget the conjoined fates of humans and nonhumans' (Kimari, 2023, p. 175). Ferdinand (2021, p. 128) thus defines the project of a *decolonial ecology* as holding 'antislavery, anticolonialism, and environmentalism together'.

Hence, he aims at bridging the gap between these struggles, with the overall goal of overcoming colonial structures, including a hierarchization of ways of living. The colonial ideology suggests, as the only true way of life, a metropolitan form of living that is based on the intensive exploitation of nature and non-humans. Ferdinand (2019, pp. 56–57) here introduces the concept of *altericide*, 'le refus de la possibilité la Terre en présence d'un autre' [the refusal of the possibility of the Earth in the presence of another]. For Ferdinand, the colonial endeavour is rendered possible by a desacralization of the Land, by the thorough destruction of a cosmogony that attributes spiritual value to the non-human.

Drawing attention to Diop's narrative is part of the project of constructing alternative ontologies of a decolonial ecology. Ibrahima Diop occupies a peri-urban space and therefore questions narratives of metropolitan life, urbanization, developmentalism and modernization – manifestations of the One-World world in Senegal.

The Senegalese city as a colonial project

From independence to the present day, Senegal, like many African countries, has lived under the reign of developmentalist thinking (Sarr, 2016). Its economic policies have been based on modernization, urbanization and industrialization, with the aim of creating continuous growth. From the 1970s onwards, these development-oriented policies have been fostered by the Bretton Woods institutions – the World Bank and the IMF. These institutions framed the countries political, cultural and social development from a neoliberal vision of the economy. The aim was to turn countries that were seen as 'traditional' or 'underdeveloped' into so-called 'modern' countries. Discourses on modernity, progress and development have considerably transformed Senegalese society. With Searle and Vandervaken (1985), we argue that these discourses have had an illocutionary force that has had cultural, environmental and social repercussions: desertification, rural exodus and disorderly urbanization. This trend has profoundly and durably relegated to the background, or even rendered obsolete or invisible, ways of living in the world that did not seem 'modern' or 'developed'. Ibrahima Diop's marginalized way of life continues to resist ideas of development and modernization in Senegal. Diop proposes a different type of modernity that might provide solutions to modern problems caused by developmentalist thinking (de Sousa Santos, 2007; Ferdinand, 2021).

Ibrahima Diop lives at the margins of the city Dakar. Today the city of Dakar is Senegal's capital. The Dakar region hosts almost a quarter of the population and 90 per cent of Senegal's industry (Cissé, 2022; Diop, 2012). On the one hand, this is remarkable since the Cap Verde peninsula on which Dakar is situated makes up only 0.3 per cent of the Senegalese territory (Cissé, 2022). Hence, there is a high concentration of the population on a relatively small stretch of land, surrounded on three sides by the ocean. On the other hand, Dakar is also a very young city. A small village in the 1990s, Dakar was established in 1857 as a French colonial city (Bigon, 2016; Diop, 2012).

As Liora Bigon (2016, p. 5) remarks, the making of Dakar 'can be regarded as an exercise in the understanding of the dynamism inherent in the colonial power relations'. The city of Dakar was established as a colonial project and to serve colonial interests. Until today, the city displays traits of a planned colonial city, an experimental terrain of French colonization (Bigon, 2016). The city, an entity of colonial creation and inspiration, remains central to the politics of the nation-state in contemporary Senegal. The segregation of the city in an African part, Medina, and a European part, Plateau, is still observable today. Dakar replaced the city Saint Louis, situated in the North of Senegal, as the capital of the Afrique occidentale française (AOF) federation in 1902 (Bigon, 2016; Diop, 2012).

Rapid urbanization transformed the village of 1,556 inhabitants in 1878 (Diop, 2012) into a metropole that is home to approximately 4 million following the population census of 2013 (New World Encyclopedia, 2024). As in other Southern contexts, Dakar's urbanization is driven by the emergence of spontaneous neighbourhoods and little centralized planning. In inhabited areas, 35 per cent feature so-called informal housing (Cissé, 2022). As Djibril Diop (2012, p. 22) remarks: '*Tout est construit, même les sites inconstructibles*' [Everything is constructed, even unconstructable sites]. The everyday life of urban dwellers is marked by traffic congestions, air pollution and noise. The Dakar region is a heavily urbanized area featuring environmental degradation.

Dwelling as resistance

This colonial background of the city of Dakar brings us to our first argument: Ibrahima Diop engages in a decolonial ecology by dwelling and surviving. Diop resists the dominant (Senegalese) narrative of modernization, urbanization and development, and thereby resists the One-World world. The ecosystem in which Diop lives is a patch of land within the city. By living and surviving, he resists urbanization. From the moment African capitals were born out of a colonial matrix, the peripheral ecosystem, as the patch of land where Diop lives, have been resisting the colonial project of a planned city.

The pressure on land in Dakar and unbridled urbanization are a threat to the preservation of biodiversity and the relational and supportive ontology that characterizes Diop's relationship with this universe and, more generally, with the surrounding community. That is why its way of life is one of those resisting the social and ecological crises arising from the capitalist development model. Every day the boundaries of the city, representing the boundary of the capitalist-industrial world of One-World imaginary and infrastructure, advance towards Ibrahima Diop's place of living. However, even though he is aware of it, he is not showing any signs of panic. He continues to live at the margins of the city. We agree with Ana Bednik (2016, p. 185) who writes:

Original Text	**English Version**
Habiter devient un acte politique, tout autant parce qu'en refusant de partir on empêche l'adversaire d'avancer [. . .] *Habiter, c'est résister à la déterritorialisation qui réduit les territoires à des fonctions économiques. C'est leur donner d'autres avenirs, mais aussi, et souvent sans attendre, concevoir et défendre leurs présents immédiats.*	Inhabiting becomes a political act, not least because refusing to leave prevents the adversary from advancing [. .] Inhabiting means resisting the deterritorialization that reduces territories to economic functions. It means giving them other futures, but also, and often without delay, designing and defending their immediate present.

The history of the city, of urbanization, is also the history of a long process of Westernization rooted in the French colonial project of *assimilation*. Ibrahima Diop is turning his back on the civilizational ideology, returning to the knowledge and practices of the local communities regarding nutrition, healthcare and transportation.

We want to draw attention to this form of resistance because it is often overlooked. While Diop's discourse and actions are, on the surface, far less vociferous and demanding than many of the militant movements we see around the world and in Africa, it is still a noteworthy form of resistance. Ibrahima Diop's survival at the margins protests the aggression of ecosystems at the outer edges of a city. We might speak of him as resisting, given the extent to which his speech, characterized by an enunciation centred on the 'hypo-local' (Garnier, 2022), ultimately proves subversive in that it challenges the hegemonic centre. This contribution aims to open up the universe and discursive imagination of resistance to ecological threats to other alternative ways of speaking and inhabiting the world, including resistance that is more than slogans. It seeks to align with approaches that move beyond a unification of the world. Hence, working at the margins of linguistics and language science allows us to resist linguistics of the One-World.

As sketched above, for Malcom Ferdinand, colonial settlement was enabled and resulted in the disappearance of the sacredness of the land. In the next paragraphs, we will first spell out Ibrahima Diop's ethics and practices of relationality. We argue that Diop, through healing and caring, is re-establishing the sacredness of that patch of land and revitalizing a cosmogony that contradicts exploitation. We see the relationship with human and non-human communities at the heart of Ibrahima Diop's resistance to the destruction of Senegalese ecologies.

Ethics of relationality and the sacrality of the other

Ibrahima Diop's ethics of relationality are rooted in a thorough understanding of '*jàmm*' [peace]: being at peace with oneself and being at peace with all life around you. In an interview with the authors, he stated:

Original Text	**English Version**
Xam nga jàmm moo neex, jàmm dafa neex [. . .] *Lépp loo gis lu nekkul jàmm rekk, man, booy dem nii danga naan yàlla nañu yendu jàmm.Jàmm moo neex quoi, jàmm dafa neex. Gisoo ñii ñëpp, jàmm laa séq ak ñoom.* [You know, peace is good, it's truly good.]	[. . .] All that you see that is not in peace, I, even when you leave me you will tell me may peace be with you. Peace is good, it is truly good. You see all these people around me; our ties are truly peaceful

'Jàmm' is a prominent expression in Wolof greetings. When encountering someone, frequently the exchange 'Mbaa yaa ngi ci jàmm?' – 'Waaw, maa ngi ci jàmm'.

['Are you at peace?' – 'Yes, I am at peace'] is uttered. Commonly, one wishes peace at departure as in 'yendul ak jàmm!' [spend a peaceful day].

However, while the word 'jàmm' is frequently encountered, the peace of mind Diop refers to is hard to find in a bustling capital. Diop explains that (only) 'àll bi' [the forest] puts you at peace. He states the following:

Original Text	**English Version**
Àll bi, li nga ci gagner mënoo ko gagner ci kër yi. Fee coow li bëri na, wax ji bëri na, histoire yi bëri na. [. . .] *Gis nga bi nga ñëwee danga maa fekk man kese ma wéet taxaw di xalaat*	What you obtain in the forest, you cannot obtain in houses. Over there, there is noise, lots of talking, many stories [conflicts]. [. . .] You saw when you came how I was, standing alone, reflecting.

Thus, only the natural environment of 'àll' [forest] offers 'jàmm' in the sense of peace of mind, while he describes life in the city as full of noise and conflict.

While the ecological potential of Southern African conceptualizations of being has been discussed widely, especially the concepts of Ubuntu and Botho (LeGrange, 2015; Ramose, 2009), much less scholarly attention has been dedicated thus far to West African conceptualization. Next to '*jàmm*', the polysemy of the Wolof verb '*taxaw*' is also interesting to reflect upon. Besides meaning to come to a halt, it also means to stand upright, and to reflect or even mediate. It is a verb that Ibrahima Diop uses often, and which becomes a symbol for his resistance. By standing and reflecting, he brings to a halt and resists the colonial effort of a developmentalist narrative and accelerated urbanization.

Urban dwellers worldwide face a range of problems, and Dakar is no exception. The past years have been marked by ever-increasing costs of living, inflation, high levels of unemployment, the lack of social policies and the slowdown in economic activity due to endogenous and exogenous economic shocks. Diop is resilient and remains relatively unaffected by political and economic events. However, that does not mean that he is completely outside of society. Diop clearly states that:

Original Text	English Version
Non, du question dëddu àddina sax, ku nekk sa vie sa dund man dund bii moo ma neex, yow dund bee moo nga neex, kee dund bee moo ko neex.	I'm not someone who leads a monastic life. I like to live the way I live, but it's not linked to a quest for an austere life. Everyone has a way of living that they prefer, you have one, someone else has one, I have one.

Instead, his way of life gives him the freedom to engage with his fellow human beings. Diop stresses that he would always set aside his current activity for any human requesting his help or attention. During a televised interview, he explained:

Original Text	English Version
Réyuma, réyuma. Dama oyof ni dàllu carax. Xale bu ndaw bu ma fekkoon fi tey mën a yàq s ama affaire yépp ma ànd ak moom.	I am humble, humble. Simple as a pair of sandals. If a child found me here today, I could suspend all my activities and follow him.

In very similar terms as above, he stated in encountering the authors: '*Yaa ngi nii ñëw rekk, maa nangu bàyyi sama yitte yëpp taxaw ak yow*' [Today you arrived unannounced, I abandoned everything to exchange with you]. Hence, Diop emphasizes the value and importance of other human beings.

However, Diop gives more than attention to others. He engages in care and healing practices based on ancestral knowledge of plants and animals, passed down within the forest communities. Within the televised interviews, he talks, for instance, about the healing powers of '*jaxatu*' [*solanum aethiopicum* or bitter tomato]. Diop states:

Original Text	English Version
Li jaxatu la, baax na lool. Boo koy lékk lu bari doo am problème bopp.	That is a bitter tomato, it does a lot of good. If you eat it a lot, it prevents headaches.

Further, Diop produces remedies and medicine on the basis of the caught snakes. He explains the healing powers of the '*graisse*' [oil] or '*diw*' [oil] of the snakes. Rubbed on different body parts, the oil can relieve pain. As he explains during the interview, he is willing to heal whoever needs him. Even though he asks for money, it is not in a capitalist logic of enrichment and expansion. Rather, it is in a logic of relationality and interdependence. Diop explains as following:

Original Text	**English Version**
Defu ma ko pour xaalis, mais, boo gisee ma naan téemeer, nit bu la defale dangay jël benn affaire rekk tég ko ci loxom, moo tax am na ñoo xam ne dañuy ñëw fi sax ma tibb suuf, jox. Ma di leen tooñ, naan leen: 'yow moomuloo suuf bi, yàlla moo moom suuf si.' Nga ma koy tuuro ma jël parce que day indiwaat client, damay indil yépp.	But I'm not doing this for money. If I ask for 500 Francs [0.80 $], it's because it's good to give something to the person who is giving you something for your pain. In fact, sometimes people who have nothing to give me, give me sand as a token. I accept, not without teasing them by saying: 'Yes, but the sand you're giving me belongs to God and not to you'. But I take the sand anyway, especially as I know it will bring me good luck and other customers.

He gives care in exchange for something, even if it is a handful of sand, which has a purely symbolic value. Next to relieving pain, the relationship that unites Diop with the people also includes rescue practices, particularly when snakes come to live in people's homes. However, he feels reluctant to take on this activity when he is called upon to go elsewhere than his neighbourhood. Diop states:

Original Text	**English Version**
Loolu de xam nga fan la koy nangu sama dëkkandoo yi ma deendal kesse kesse ñoo ko mëna am ci man. Wax naa la loolu nekkul sama métier.	These days I only accept to remove snakes if my neighbours solicit me. I told you, it is not my occupation.

Hence, next to relations with humans, Diop also advocates relating to snakes. Diop talks about snakes not as dangerous others but as beings to relate to. For instance, he says:

Original Text	**English Version**
Moo ngi nii, xoolal, yeew bi, mujje na nekk, sax sama xarit. [. . .] *Maa ko baña déranger mais buma duggee, naa ko jàpp.*	It is here, you see, the snake; it is behind there, because it is my friend. [. . .] I refused to disturb it, but if I entered, I catch it.

For snakes, he never uses the verb '*rëbb*' [to hunt] but speaks about '*jàpp jaan*' [to catch snakes] and '*fowe*' [to play with]. Thus, his aim is not to kill the snakes, but he has a certain knowledge that allows him to relate to the snakes. At times, people aware of his knowledge call him to catch a snake, but he relates to the snakes, plays with them and liberates them afterwards. Hence, in Diop's utterances and actions, we see a practice of the relational, of the absence of a division of nature/culture, human/non-human or individual/community (Escobar, 2018). Diop underlines the value of others, be it snakes or unannounced human visitors, for his way of life.

His idea of a coexistence with snakes is of importance in trying to understand an event that received much media attention. In a televised interview, Diop explained that an elder convinced him to (spiritually) coach the wrestler Ama Baldé, even though he had no prior experience in wrestling. That is how the wrestler once appeared in the stadium with a python on his shoulders that was supposed to frighten his opponent. In Senegal, wrestling fights are accompanied by dancing, griot songs and a range of mystical preparations including talismans and anointing oneself with protective liquids called 'saafara'. The popular belief is that to win a fight, the wrestler needs not only physical but also spiritual qualities or magical protection (Bonhomme & Gabail, 2018). Diop, in the interview, explains why he refuses requests from other wrestlers to help them carry a snake or python. He explains that God would condemn an act in which a snake is simply used to make money. For Diop, the peaceful coexistence with snakes is a supernatural gift that should not be sold out. Indeed, since the 1990s, wrestling in Senegal has become a very lucrative business that allows wrestlers and managers to earn millions. It is said to be the best-paid professional sport in the entire African continent (Bonhomme & Gabail, 2018).

In line with the 'payment' with sand discussed above, Diop is part of a money-making economy. While his utterances and acts can be interpreted as resisting capitalist modernity, he is nonetheless mutually entangled in it, and at times 'takes sides'. Therefore, Ibrahima Diop's self-narration and hetero-perception are to an extent ambiguous.

Limits of interaction and the role of new media in rediscovering a decolonial way of life

The journalist of Pexe TV introduces the interview as an opportunity to get to know Iran '*dëgg-dëgg*', the 'real' Iran. While Ibrahima Diop, better known by his short name Iran, is today a media persona in Senegal, he has not always been perceived positively by the public. The necessity to present the 'real Iran' also means countering the negative prejudices associated with Ibrahima Diop. The journalist states that Iran is commonly referred to by the wider population as '*Iran nekkul nit ku jub; Iran nekkul nit ku baax*', which translates to English as 'Iran is not a righteous person; Iran is not a good person'.

Often, all those living in the peri-urban forests of Dakar are referred to as 'bandits'. The word 'bandit' is commonly used in urban Wolof to designate both a crook who makes a living from burglary and petty theft, and the 'deviant' who, without engaging in banditry, is nonetheless likely to frequent 'dangerous' places: dense forests, groves, isolated beaches, etc. There is a focus in Senegalese social sciences on organized crime, large-scale robbery and white-collar banditry, which is the work of predatory elites (Cissé, 2022; Daha, 2002; Faye, 2021). However, they have paid little attention to the social figure of the 'bandit' from the backwoods, who is a 'bandit' in name only, not invested in crime. But in the words of the journalist who interviewed him, Iran has long been perceived by the local population as a 'bandit'. His solitary life in the small woods, his excursions from one point of the bush to another in the company of a dozen dogs, his unkempt look – in short, his 'marginal' appearance – played a large part in shaping this perception. Iran thus devotes part of the interview to deconstructing this hetero-perception in order to highlight the ethos of an individual: close to nature and animals, disinterested, humble and altruistic. On the one hand, his auto-presentation as being far removed from the quest for profit and visibility, and the claim to be rooted in values of humility and humanism, diverges from the hetero-presentation of 'banditry'. On the other hand, it points to Iran's lived and practiced relational ethics, which is very much in line with theoretical reflections of Arturo Escobar.

Ibrahima Diop for a long time refused to be interviewed. When he finally decided to present himself, a new narration emerged. Through interviews and social media posting, one gains an almost intimate understanding of him. The videos of Diop offer to the public everyday gestures, such as biting into a raw aubergine, playing with an animal and growing vegetables: symbols of an auto-representation displaying humility, playfulness and selflessness.

Documenting ecological destruction in contemporary Senegal

Above, we quoted Diop in praising '*àll bi*' [the forest] in giving peace of mind. However, the definition of *àll* diverges from the English 'forest'. Following

Diouf (2003), the possible translations of *àll* include a non-inhabited place, the bush, landscape and nature. In contemporary Senegal, the lexeme *àll* is commonly used to denote everything that is not fully urbanized, from grasslands to stretches of sand. Hence, in Diop's utterances, *àll* is an antonym to Dakar.

During interviews, Diop maps the *àll* of Dakar and creates an inventory of the wild animals living there. These include 'bar' [Nil monitor], 'lapin' [hare], 'xojog' [sun squirrel] and '*mbëtt*' [savannah monitor]. Diop even provides a list of other 'marginals' that could be found in these woods. To some extent, the lexicon he uses here seems to be in conflict with the reality he describes. He refers to urban realities with words such as 'sayyaan' [water place] and 'rëbb' [hunt].

According to him, the paradise that *àll* represented is no more. The fauna and flora have decreased as a result of rampant urbanization. Diop spells out the idea of a delightful past in 'àllbi' in stating:

Original Text	**English Version**
Fi amoon na àll te fi amatul àll fi, xam nga, lan moo fi am jardin. Bu njëkk ba la ko daale fi ba autoroute ba yépp ay garab la woon, ay tool la woon, dex bi nekku fi woon. Dex bi nga gis ni nekku fi woon, daa doon féer danga fi doon joge jubbal ba Sandica ay tool la woon mais li ñu massale tool yi loolu moo tax kenn rëbbatu fi mais deñ doon rëbb.	There used to be *àll* here, but it does not exist any longer! You know what is it now? A garden! In the past, from here to the motorway, there were nothing but trees and fields, and the pond you see there wasn't there. It dried up after the rainy season. From here to the Syndicat market in Pikine, there were only fields. But now that these fields have been cleared, we can no longer go hunting, which we used to be able to do.

Ibrahima Diop's relationship with his convironment should not be mistaken for primitivism or nostalgia for an African Eden. Nowhere in his interviews do we find the fantastical aspects of these tropes. Nor is there any rhetoric of mystification or desolation. Diop's approach is fully practical. He encourages us to experience the environment. In this respect and following Garnier (2022, p. 13), we could credit him with a decolonial stance:

Original Text	**English Version**
L'abord fantasmatique des lieux africains est donc une affaire de conquérants. Du point de vue des colonisés, les lieux du continent sont pratiqués avant d'être rêvés, et non l'inverse. Ce sont des lieux a priori peu concernés par les mythologies et les cartographies dans lesquelles on les insère	The phantasmatic approach to African places is a matter for conquerors. From the point of view of the colonized, the continent's places are experienced before they are dreamt of, and not the other way round. The mythologies and cartographies into which they are inserted have little to do with them

Diop laments that the 'modern' vision of Senegal gradually destroys what is left of *àll*.

Original Text	**English Version**
Léegi loolu lu mu mun a def ci ab réew, ci ab capitale, ñu dem ba nga ba ci biir hum ñun nga xam ne. Ay africains lañu hum dañu mos di dund ci biir ci wetu àll walla xëy na sax ci biir àll. Ñu dem ba ci bir léegi di dëkk foo xam ni ay ciment rekk a fa nekk ak ay tali, ñu dem ba nga xam ne léegi àll amatul jardin rekk a fa des, ñu dem bay xalaat naan jardin de boo moytuwul sax bu ñu demee ba ci biir dina jeex. Dëkkuwaay yi dañuy yokk, bu njëkk, léegi dina dem ba nga xamante pour nga gis àll rek dangay génn. Xale yi mujj duñu xam dara, xëy na dinañ xam ay affaire portable xëy na ak di gis ay kii nii, mais ñun li ñu dund ñiy ñëw ak ñi si topp munuñ ko dund.	It can transform a country, a capital. We Africans who used to live in or near such environments see them destroyed. We are so obsessed with cement, roads . . . that we no longer have forests but only gardens. And maybe at some point we won't even have gardens anymore. Habitats are constantly expanding. If this continues, you'll have to travel to see a forest. Children will know nothing about forests. Maybe they'll be able to see what a forest is through mobile phones, but they won't be able to experience what we have here, because as I've told you, habitats are progressing.

On the one hand, *àll* is endangered by large-scale governmental projects. As written on the presidential website, 'Senegal has decided to adopt a new development model to speed up its march towards emergence. This strategy, named Plan for an Emerging Senegal (PES), forms the reference framework for the country's economic and social policy over the mid- and long-term.' (Presidency of Senegal, n.d.). Launched in 2014, this plan aims at 'developing'

Senegal with projects that very much follow a colonial ideology of exploitation of nature. Among key projects are gold mining, iron mining and drilling of gas. Furthermore, a new city or 'urban pole' is currently being constructed about 30 kilometres outside the Dakar city centre. Hence, the not yet urbanized spaces at the outskirts of Dakar are endangered and will be destroyed in the upcoming years.

On the other hand, *àll* is endangered by small-scale human action. Diop talks about deforestation by individuals living close by. Since talking to these individuals proved unsuccessful, Diop started to call the state institution in charge to 'defend' the forest:

Original Text

Boo gisoon fee nit di dagg garab bëgg ma kii ma wax ko mu dëgër bopp, ma woo eaux et forêt yi ne kii mu ngi dagg garab yi, eaux et forêt yi dinañ la fi fekk dagg garab yi, keneen sama benn morom def noonu tamit, noonu lay demee rekk. Parce que bo bàyyegars yi dañuy dagg garab yi garab yi jeex tàq. Xam nga ku nekk ak li nga soxla. Mun nga am benn soxla garab bee yow nga ñëw dagg ko yóbbu.Am na ñoo xam ne dañuy ñëw.

English Version

You can see people cutting down trees and being stubborn when they're asked not to, I report it to the forestry officials. Because if we don't do anything . . . people will cut down all the trees. They need the plants for wood and various other uses, including care. Some people only come for wood, charcoal . . . Because if it's for treatment, the leaves can do the trick. But they cut down trees mainly for wood. In the long term, this is a threat that could eradicate the forest.

In the way of life of Diop, resisting deforestation is part of the everyday. Like other activists, he faces environmental threats on a daily basis. The relationship with human and non-human communities is at the heart of his resistance to deforestation. Instead of devouring the world as exploitative practices do (Ferdinand, 2019), Diop lives a self-sustainable life on the margins of the city. Diop's primary protest is to live and be. However, his existence is increasingly being threatened. Thus, on a secondary level, he engages with the media and the forestry office. While calling the forestry department is a means of ensuring his marginalized way of life, this collaboration also adds a layer of ambiguity. To an extent, relying on this state institution puts into question the reading of Diop's narration as decolonial.

The limitations of a decolonial ecology

As stated in the introduction, we advocate an inclusion of marginal biographies and voices in ecolinguistics. We believe that this is a necessary step towards decolonizing ecolinguistics as a discipline. However, instead of being a monolith, the margin is a space of struggle, contestation and ambiguity.

This ambiguity surfaces most clearly when Diop talks about calling the forestry officers. In practice, it means that the 'marginal' Diop collaborates with an environmental police force to defend his claims against other marginalized people of Dakar. While the forestry institution certainly prevents trees from being cut down, it has not completed its decolonization for a number of reasons. First, the institution has been criticized for acting in disassociation with the communities it encounters. It tends to consider the communities as irresponsible and wants to educate them or raise their awareness. Instead of being willing to learn from local ways of living and acknowledging the sacredness of 'àll', the forestry office pursues a classic colonial approach to environmental protection. Second, in its work, punishment and repression are all too often its modus operandi. The officers come in uniform and with weapons to monitor and punish, and do not shy away from resorting to violence. This is in complete contradiction to the relational ethics sketched above. A related criticism that could be levelled at Diop is his dedicated silence regarding the authorities responsible for political, economic and social measures of deforestation. Individuals cutting wood do so for their survival; they are also part of the precarious population trying to survive in post-colonial Senegal. Hence, by cooperating with local forces, Diop might be endangering the very foundations of his way of life.

However, in concluding this reflection, we come to understand that the exploration of the margins complicates the assumption of departure of the decolonial endeavour. As sketched in the beginning, at the heart of decolonial critique lies the assumption of an imbalance between centre and margin. Our chapter complicates the divisive lines at several stages. As we have shown, North and South are not necessarily the antagonists in Ibrahim Diop's biography. We are observing a nuanced network of survival struggles within Senegal. These are embedded in a global economy and reflect a violent history of colonization. However, it is also necessary to highlight that neither contemporary Senegal per se nor (urban) Wolof as such offers a decolonial ecology. Taking seriously the claim of thinking the world from the (African) margins means to be very well informed about the societal tensions within the South.

Our chapter shows how difficult it is to uphold the differentiation of North and South or colonizer and colonized. Diop's way of life, which is respectful to his convironment is endangered in at least three ways. First, large-scale infrastructural projects and modernization aspirations of the post-colonial Senegalese state endanger the remaining peri-urban forests. Second, those Senegalese living at the margins threaten to destroy their own livelihoods in their quest to aspire to integrate into an ever-growing city, promising modernization. Finally, Ibrahima Diop is a threat to himself and his way of life, for instance by choosing allies that adhere to a modernization agenda. Even more important is to draw inspiration from his marginal narrative to change realities in the centres of this world.

Note

1 We privilege the term convironment over environment because it emphasizes the reciprocal relationships and mutual constitution of the world.

2 In Dakar, most people use urban Wolof, a variety of Wolof characterized by extensive mixing of Wolof and French lexicon and grammar.

References

Bednik, A. (2016). *Extractivisme. Exploitation industrielle de la nature : logiques, conséquences, résistances.* Le passager clandestin.

Bigon, L. (Ed.). (2016). *Place names in Africa. Colonial urban legacies, entangled histories.* Springer International Publishing.

Bonhomme, J., & Gabail, L. (2018). Lutte mystique. Sport, magie et sorcellerie au Sénégal. *Cahiers d'Études africaines.* Retrieved from https://doi.org/10.4000/etudesafricaines.22834

Cissé, M. (2005). Les politiques linguistiques du Sénégal: entre attentisme et interventionnisme. *Kotoba to Shakai [Language and Society]: Special issue on post-empire and multilingual societies in Asia and Africa*, 266–313.

Cissé, O. (2022). *Challenges of urbanisation in Dakar. Land-use planning, sanitation, public transportation and public housing.* Friedrich-Ebert-Stiftung.

Daha, Ch. (2002). *Marginalité et exclusion au Sénégal: les comportements delictuels et criminels dans la vallée du fleuve Sénégal (1810–1970).* [Doctoral dissertation, Université Cheikh Anta Diop de Dakar].

De Sousa Santos, B. (2007). *Democratizing democracy. Beyond the liberal democratic canon.* Verso.

Diop, D. (2012). *Urbanisation et gestion du foncier urbain à Dakar. Défis et perspectives.* L'Harmattan.

Diouf, J.-L. (2003). *Dictionnaire wolof-français et français-wolof.* Karthala.

Diouf, I., Ndiaye, Ch. T., & Dieme, N. B. (2017). Dynamique et transmission linguistique au Sénégal au cours des 25 dernières années. *Cahiers québécois de démographie, 46*(2), 197–217. https://doi.org/10.7202/1054052ar

Escobar, A. (2016). Thinking-feeling with the Earth: Territorial struggles and the ontological dimension of the epistemologies of the South. *AIBR, 11(1)*, 11–32. https://doi.org/ 10.11156/aibr.110102e

Escobar, A. (2018). *Sentir-penser la terre. Une écologie au-delà de l'Occident.* Seuil.

Faye, Ch. (2021). Water resources and their management in an increasing urban demography: The case of Dakar City in Senegal. In P. Th. Chandrasekaran, M. S. Javaid, & A. Sadiq (Eds.), *Resources of water*. IntechOpen.

Faye, O., & Thioub, I. (2003). Les marginaux et l'État à Dakar. *Le Mouvement Social, 3*(204), 93–108.

Ferdinand, M. (2019). *Une écologie décoloniale. Penser l'écologie depuis le monde caribéen.* Seuil.

Ferdinand, M. (2021). *Decolonial ecology: Thinking from the Caribbean world.* John Wiley & Sons.

Garnier, Xavier. (2022). *Écopoétiques africaines. Une expérience décoloniale des lieux.* Karthala.

Kimari, W. (2023). Book Review. Ferdinand, Malcom. 2021. Decolonial Ecology: Thinking from the Caribbean World. *Environment and Society: Advances in Research, 14*, 175–177.

LeGrange, L. (2015). Ubuntu/Botho as ecophilosophy and ecosophy. *Journal of Human Ecology, 49*(3), 301–308.

Ly, M. A., & Rohmer, M. C. (2024). Présentation du Numéro des Notes Africaines sur : Regards *interdisciplinaires* sur la recherche écolinguistique au Sénégal. *Notes Africaines.* (tpb)

McLaughlin, F. (2008). Senegal: The emergence of a national lingua franca. In A. Simpson (Ed.), *Language and National Identity in Africa* (pp. 79–97). Oxford University Press.

McLaughlin, F. (2017). The social meanings of Wolof and French: Contact dialects, language ideology, and competing modernities in Senegal. In R. Bassiouney (Ed.), *Dialect and identity performance* (pp. 179–191). Routledge.

New World Encyclopedia. (2024, Jan 24). *Dakar*. Retrieved from https://www.newworldencyclopedia.org/entry/Dakar

Presidency of Senegal. (n.d.). *Emerging Senegal.* Available online: https://www.presidence.sn/en/pse. Accessed July 7, 2022.

Ramose, M. B. (2009). Ecology through Ubuntu. In M. F. Murove (Ed.), *African ethics: An anthology for comparative and applied ethics* (pp. 308–331). University of Kwazulu-Natal Press.

Sarr, F. (2016). *Afrotopia*. Philippe Rey.
Sarr, F. (2017). *Habiter le monde. Essai de politique relationnelle.* Mémoire d'encrier.
Searle, John R., & Vanderveken, D. (1985). *Foundations of Illocutionary Logic.* Cambridge University Press.

10

An ecolinguistic analysis of the role of civets in Indonesian 'civet coffee'

Ingrid A. Gavilán Tatin, George M. Jacobs, Meng Huat Chau, Chenghao Zhu and Aditya Alam Soeta Bangsa

Introduction

Kopi luwak or civet coffee is the Indonesian term for a coffee drink made from coffee beans that have been excreted by civets. This introduction provides background on civets and on coffee production in Indonesia. The focus of this chapter involves an exploration of the representation of civets and civet coffee in two leading Indonesian newspapers. Analyses were performed using tools and insights from ecolinguistics, which can be defined as an examination of 'the role of language in the life-sustaining interactions of humans, other species and the physical environment' (International Ecolinguistics Association n.d., para 1).

Background on civets

Civets are mostly nocturnal, omnivorous mammals who live in tropical regions of Africa and Asia. More than ten species of civets exist, and civets have various names including civets and *toddy cats* in English and *musang* in Malay (National Parks Board, 2022; World Animal Protection, n.d.). In Indonesian languages, civets are referred to by various names, such as *luwak* in Javanese, *careuh* in Sundanese, and *musang*, *ferrets*, and *mongooses* in different regions, with modifiers added to *musang* to specify different civet species (Cahill, 2019). The current study focuses on the civet species known as the palm civet. They are native to South and Southeast Asia and are smaller than some other civet species, varying in body length from 43 to 71 centimetres, with a tail length of 41 to 66 centimetres. Their weight varies between two and five kilograms. Palm civets are grey in colour with black markings and bodies resembling those of cats. That said, other animals, including binturong and even elephants (Jamal Said, 2022; Sullivan, 2014), have been used to produce coffee via a similar process. Fridayani (2020) claimed that Black Ivory Coffee, in which coffee beans are extracted from the faeces of Thai elephants, is even more expensive than civet coffee. Civet coffee is also produced in the Philippines and perhaps other countries.

Civet–human interaction has taken various forms. Civets have been hunted in the wild for meat, as well as in a form of pest control, as civets eat farmers' crops and animals. While the current chapter discusses how humans use civets as part of coffee production, civets have also been used in the production of perfume. Additionally, during the SARS virus pandemic, civets were believed by some to be involved in the spread of SARS (Roos, 2004), and, as a result, some civets were culled by governments in hopes of reducing the impact of the epidemic (Parry, 2004). Ivory (2021) talked about civets and related animals as pets. Among the advice was to keep them in cages but allowing some time out of the cage, at least to be able to clean the cage. Although Ivory linked interest in civets as pets to the topic of civet coffee, feeding coffee beans to civets was not mentioned.

Coffee production in Indonesia and the role of civets

The use of civets in commercial coffee production is relatively new. Coffee itself was introduced to Indonesia in the late seventeenth century by the Netherlands, the major colonial power in Indonesia. Using Batavia (modern-day Jakarta) as their main hub, the Dutch monopolized maritime trade routes during this time through the Vereenigde Oost Indische Compagnie (Dutch East India Company) (Barrow, 2017). Along with other commodities, coffee rose in popularity in European markets, which encouraged its production in Indonesia. Indonesia became an independent country in 1945, and today, coffee production is a major engine of Indonesia's economy. According to Deshmukh (2021), Indonesia ranks as the world's fourth-largest coffee-producing country. In Indonesia, more than 90 per cent of coffee is produced by smallholders, with more than 1.7 million people employed (World Coffee Research, 2023). It should be noted that civet coffee accounts for a very small proportion of Indonesia's coffee production and is seen as a novelty item, although one which has managed to attract a great deal of attention.

According to Cahill (2019), civet coffee arose during the Dutch occupation as coffee producers gathered the beans from civet faeces found near coffee plantations. This alternative method allowed locals to enjoy coffee despite the prohibitive policies of the time that reserved coffee for use by the Dutch and their favoured Indonesians. Government interest in civet coffee became significant in 2006, with this type of coffee garnering high prices. Of the ten foods classified as most expensive (Langit Nariswari, 2023), seven, including civet coffee, involve the use of non-human animals. A further sign of government interest was that civet coffee was given *halal* certification (certified appropriate for the diets of Muslims) in 2010. More than 90 per cent of Indonesians identify as Muslim.

As with many products, buyers need to beware, as claims of the authenticity of civet coffee and whether it comes from wild civets may be false (Agmarasi, 2016). Although civet coffee from wild civets is believed to be superior to that from captive civets, to date, no certification system exists, and a 2022 report by PETA (People for the Ethical Treatment of Animals) claimed that civet coffee labelled as wild-sourced actually came from caged civets. Furthermore, coffee sold as civet coffee may be a mix of regular coffee and civet coffee or contain no civet coffee at all.

Concerns about the treatment of civets

Originally, civet coffee production relied on the natural collection of civet faeces in coffee plantations and nearby forests. However, in an effort to increase civet coffee production, some civets were caged and fed coffee beans as their main diet. Overseas media and animal welfare organizations, such as the BBC's Lynn and Rogers (2013) and PETA (2022) highlighted what they saw as abuse due to the caging of the civets. In a commentary in *The Jakarta Post* (a prominent Indonesian English-language newspaper) responding to a BBC exposé about the treatment of civets in the production of civet coffee, Peresthu (2013) compared the treatment of civets with that of other non-human animals whom humans use to produce other foods and beverages, concluding that the mistreatment was the same and that the overall practice originated from a Western commodity production model.

Some writers raised the possibility of an alternative to caging civets in captivity in order to produce civet coffee. For example, Peresthu (2013), a chef and restaurateur, claimed that connoisseurs prefer civet coffee from wild civets to the same product from caged civets. While acknowledging that switching entirely to wild-sourced civet coffee would result in decreased supply, Peresthu claimed that using wild-sourced beans was preferred not just for animal welfare reasons but also because it could provide small farmers with a significant source of supplementary income, as well as offering a more equitable arrangement between farmers and intermediaries. Indeed, this important role for small farmers does seem to be the arrangement in some areas of Indonesia (Djumena, 2013; Rachmawati, 2016).

In contrast to claims of mistreatment of caged civets, several articles argued that caging civets did not harm them. For instance, Nursastri, 2017 talked about the civets being treated as members of the farmers' family and eating nutritious food rather than being fed only coffee cherries (i.e. beans). Furthermore, such dedicated care of the civets was given as one of the reasons for the relatively high price of civet coffee.

Methods

Articles from two of Indonesia's leading newspapers, *Kompas* and *The Jakarta Post*, were collected from April 2008 to September 2023 pertaining to civet coffee. The Selenium WebDriver, a popular open-source tool for automating web browsers in Python, was employed to access the websites of the two newspapers. It also enabled the use of the search functionalities of the respective newspaper websites. The search queries in this study included the terms 'kopi luwak', 'luwak coffee' and 'civet coffee.' Through this approach, a substantial corpus of articles was developed in line with the prescribed search terms, which ensured their relevance to the research topic of this study. Given that our data sources include articles in both Bahasa Indonesia and English, this study utilized the integrated translation functionalities of the websites to translate those articles in Bahasa Indonesia to English.

Three of the researchers of this study manually checked the collected articles and excluded eight cases from the analysis. These were cases in which key terms such as 'kopi luwak' were mentioned exclusively within the 'Baca juga [Also read]' section, as follows:

Baca juga: Termahal di Dunia, Ini Sejarah Kopi Luwak di Nusantara [*Also read: The most expensive in the world, this is the history of Kopi Luwak in the archipelago*].

Based on this manual analysis, 170 articles from *Kompas* and 24 articles from *The Jakarta Post* were identified as relevant and stored in an Excel file for subsequent analysis. Furthermore, a distinct identifier was allocated to each newspaper article. To illustrate, the identifier 'KP001, 2008' designates the first article collected from *Kompas*, which was published in 2008, while 'JP001, 2013' corresponds to the first article collected from *The Jakarta Post*, which was published in 2013.

Much of the ecolinguistics analysis was based on the work of Stibbe (2021), who identified nine types of stories in texts. These stories influence people's thoughts and actions. A detailed summary of these nine stories – ideology, narratives, framing, metaphor, evaluation, identity, conviction, erasure and salience – is beyond the scope of this chapter. For more details, readers are kindly referred to a free online course offered by the International Ecolinguistics Association. In the current chapter, four types of stories are examined: framing, ideology, metaphor and erasure.

Results/findings

In this section of the chapter, the results of qualitative analyses of the newspaper data are presented and discussed. This is ordered according to the following topics: changed perception of civets; wild civets vs. captive civets; caged civets in tourism; production process; civet coffee: taste, quality and price; civets as machines and workers; and the farming of non-human animals.

Changed perception of civets

The popularity of civets grew due to their utility in the coffee market, making them essential assets for the production of civet coffee. This highlights how societal perceptions and economic interests can influence the status of non-human animal species. Another example of how changing perceptions of civets can lead to drastic consequences occurred in 2004, when approximately 10,000 civets in Guangdong, China, were exterminated, as it was believed they transmitted SARS to humans (Bliss, 2017). Similar killings occurred elsewhere.

1. 'Yes, even though some time ago luwak coffee was perceived as dung, now, it's the opposite. This coffee is sought after at a high price, the price even reaches IDR750 thousand per kilogram,' [as of October 2023, one US dollar equaled about 16,000 Indonesian rupiah], said M Sapri [Luwak coffee entrepreneur from West Lampung]. (KP010, 2010)

2. In the past, Indonesian people considered civets to be annoying pests. Farmers feel disadvantaged by ferrets, because the animals destroy agricultural fields and eat the fruit and vegetables they plant. But that was before. Now, people are looking at ferrets or mongooses as pets at home. The trend of keeping ferrets emerged at the same time as the civet coffee trend that occurred several years ago. (KP129, 2021)
3. The civet cat was in fact a creepy pest scurrying over rooftops and eating prize-winning tajen cocks. Their utility hadn't been explored at all as coffee 'fermenters' and their fecal matter was a mere inconvenience to the villager, as is dog crap to the jogger in New York City. (JP005, 2015)

The previous examples correspond to what Stibbe (2021) referred to as framing, or how we organize our views of the world in our minds. Hameed (2021) elaborated on this by pointing out that people first create stories using language, and as these stories gain popularity, narratives on related topics merge to become frames. These frames serve people as guides for comprehending the world in which they live.

The framing of civets changed with the recognition of their potential as a financial resource. Narratives of civets' skill in choosing the ripest coffee cherries and the exceptional quality of coffee beans found in their faeces have led to civets being closely associated with deluxe coffee in Indonesia and with status for the country as a place to find unique coffee products. This has brought about a transformation in the way civets are perceived (Cahill, 2019).

Wild civets vs. captive civets

The two methods for harvesting coffee beans after they have passed through the digestive systems of civets are wild civet coffee production, also known as the natural or traditional method, and the caged method, in which civets are captured in the wild and confined to cages. Carder et al. (2016) and D'Cruze et al. (2014) argued that gathering beans from the faeces of wild civets may offer a mutually beneficial approach that does not pose a threat to civet welfare. This approach only allows people to profit from non-human animals who might otherwise be considered nuisances due to their consumption of coffee beans and other commercial products, such as chickens' eggs, and, in this way, fosters coexistence between humans and civets. From a consumer perspective, this traditional method is believed to produce a higher-quality product, further emphasizing the potential for sustainable and ethical civet coffee production. This might be somewhat equivalent to families keeping cows or hens as pets and consuming or selling the milk or eggs, respectively. Obviously, such production methods are very small-scale and would do little to boost a country's economic output or even a family's income.

In the case of captive civets, there is a very clear anthropocentric perspective based on the benefits humans can gain from non-human animals. The main reason for confining civets is to increase production levels (Muzaifa et al., 2019). Nevertheless, this production method affects civets in many ways; for example Carder et al. (2016) reported that some civets were kept in cages that afforded them little room for movement, adding that in most cases there was no provision for natural vegetation or environmental enhancement in the cages and that hygiene was poor. This situation is similar to that faced by egg-laying hens, who often are confined, about four hens to a cage, without room to even spread their wings or perform other natural behaviours, such as dust bathing (Koknaroglu & Akunal, 2013). An Indonesian Ministry of Agriculture regulation (2015) allows the use of caged civets for civet coffee production for a maximum of five years of confinement for an individual civet. After this period, civets should be released into their natural habitat or kept as pets. However, freeing civets does not prevent them from being recaptured (Hooper et al., 2022).

Animal welfare practices vary from one country to another due to factors such as culinary traditions, economic development, education and public perception (Koknaroglu & Akunal, 2013). Sollund (2011) built upon Agnew's (1998) ideas and offered three situations in which non-human animals are mistreated. The first situation occurs when people are unaware of the negative consequences of their treatment of animals. The second is when individuals do not acknowledge that their treatment of animals is morally wrong. The third situation may overlap with one of the first two, that is, when individuals derive personal benefits or perceived gains from their acts of cruelty towards other animals. This aligns with Regan's (2003, p. 12) assertion that 'animal agriculture is a business, after all, whose object is to maximize financial return while minimizing financial investment.' This implies that the primary focus in such cases is on financial profit rather than animal welfare. Moreover, in Indonesia, there is a prevalent cultural practice of keeping animals in cages, which has become normalized. This practice includes a wide range of animals, such as dogs (Keil, 2016) and birds (Harris et al., 2017; Iskandar et al., 2019; Nijman & Nekaris, 2017) even though they are considered pets.

Bliss (2017) reported that the coffee industry faces pressure to address concerns about counterfeit civet coffee labelled as 'wild sourced.' Technology has been used to differentiate civet coffee from other coffee, and coffee from wild civets from that from caged civets. Wakhid et al. (2020) used sensors to distinguish civet coffee from non-civet coffee and reported that 'civet coffee is often falsified with non-civet coffee' (p. 56). Furthermore, Prajna et al. (2023) reported using technology to distinguish wild civet coffee from caged civet coffee, although they did not report using their technology to test the veracity of producers' claims about the circumstances of the civets that they used. Lachenmeier and Schwarz (2021) speculated that one reason for the differences between the coffee from wild and caged civets may be civets' preference for a particular type of coffee, a preference which they are unable to exercise when caged. Below are excerpts from the two newspapers which discuss wild and caged civets.

Wild civets

4. This established business model [wild civet coffee production] involves picking up civet droppings in the wild, living harmoniously with nature and respecting the animal welfare. Compared to farming civets, the traditional way is cheaper for the farmer. They do not need to purchase the wild civets, build the cages and feed them on a daily basis. Harvesting wild luwak coffee is free for the farmers. Wild luwak are always available every day in nature as long as they [agricultural workers] are willing to walk far and work hard to visit the peripheries of nearby preserved rainforest. (JP002, 2013)
5. 'Civet coffee from wild civets has a richer taste because wild civets use their instincts to find the best coffee beans. From a business perspective, captive civets are usually fed coffee beans. The price of wild civet coffee is of course more expensive because the effort required is greater,' said Aris [Indonesia Coffee Academy's trainer]. (KP082, 2016)
6. The price of luwak coffee will be even more expensive if this coffee is free range, aka coffee obtained from civets that are allowed to live wild and choose their own coffee cherries. (KP154, 2022)

Caged civets

7. 'About 10–30 per cent of the bred mongooses die or escape, which will definitely affect its population,' said Guntoro [a researcher from Assessment Institute for Agricultural Technology Studies (BPTP) Bali]. (KP009, 2010)
8. Ferrets should also be kept in cages individually because they tend to attack each other. If a ferret is injured, the wound usually rarely heals so it can end in death. (KP023, 2011)
9. Civets in cages do not move freely to choose the best coffee cherries. The mongoose just remained silent, while the keeper offered him food. 'In fact, civets have the instinct to choose the ripest coffee from the tree,' said Jamin [a coffee farmer in Bali]. And this instinct is the initial process to produce the best quality luwak coffee. (KP030, 2011)

Stibbe (2021, p. 2) stated that, 'It is through language that the natural world is mentally reduced to objects or resources to be exploited and it is through language that people can be encouraged to respect and care for the systems that support life.' In the context of the excerpts related to wild civet coffee, while there are hints of beneficial discourse with ideologies aligning with ecological and ethical beliefs, the predominant story is what Stibbe has called evaluation. This evaluation involves the stories in people's minds about whether this area of life is good or bad (Stibbe, 2021).

The text reflects a series of positive evaluations, underscoring economic advantages, through the use of specific words and phrases. For example, the word '*cheaper*' as found in sentence (4), 'Compared to farming civets, the traditional way is cheaper for the farmer', highlights how cost-effective the conventional way of producing coffee is. The phrase '*do not need to purchase*', as found in sentence (4), 'They do not need to purchase the wild civets, build the cages and feed them on a daily basis', has a positive connotation by reducing the cost of production. In sentence (5), 'Civet coffee from wild civets has a richer taste because wild civets use their instincts to find the best coffee beans', '*richer taste*' is used to emphasize the superior quality of the final product, and in sentence (6), 'The price of luwak coffee will be even more expensive if this coffee is free range', the phrase '*even more* expensive' is used to allude to the potential for increased income. All of these positive appraisals are related to the economic aspects of coffee production but do not directly concern the treatment or welfare of non-human animals.

In the excerpts related to caged civets, a central anthropocentric (human-centred) perspective is evident. The story of the evaluation of civets in these texts is primarily expressed through the use of negative appraisals, emphasizing human-centric concerns over animal welfare. In excerpt (7), 'About 10–30 percent of the bred mongooses die or escape, which will definitely affect its population', the phrase '*die or escape*' highlights the potential negative outcomes for bred civets, framing the issue in terms of its consequences for human interests, specifically civet population management, while overlooking the well-being of the animals themselves. Similarly, in excerpt (8), the term '*injured*' brings attention to the potential harm suffered by the non-human animals, as can be seen in the sentence 'If a ferret is injured, the wound usually rarely heals so it can end in death.' However, the main concern is presented as the inconvenience or consequences for humans, such as the death of the animals who would need to be replaced, rather than a primary focus on the animals' welfare. In excerpt (9), 'Civets in cages do not move freely to choose the best coffee cherries', the phrase '*do not move freely*' underscores the restriction of civets' natural behaviour in captivity. This limitation is framed as problematic due to its impact on the quality of coffee production, rather than addressing ethical concerns related to confining fellow animals.

Whether the appraisal is positive or negative, the excerpts from the two newspapers collectively reflect an anthropocentric ideology where non-human animals are viewed primarily through a lens of utility to and convenience for humans, with the well-being of the civets taking a secondary role to human animals' interests.

Defenders of humans' current treatment of non-human animals used to produce food for humans sometimes claim that cows, turkeys, fishes etc. are just 'dumb' animals, incapable of suffering. However, Singer (2015) argued that the innate capacity for physical and emotional pain is inherent in many non-human animals. Singer emphasized that humans' fellow animals should be recognized as sentient creatures with complex inner experiences. In other words, these farmed animals are beings who possess the ability to think, feel and desire (Zhdanava et al., 2021).

Caged civets in tourism

A number of articles in the database highlighted the role of the civets themselves as tourist attractions, rather than the coffee that the civets help produce. Another case of the use of non-human animals' eating habits as a tourist attraction occurs in Morocco, where tourists travel to view goats climbing trees to eat nuts (Marocknroll, n.d.). Putting civets in cages, similar to putting other wild animals in zoos and circuses, makes it easy for tourists to view them. This practice may be another example of anthropocentrism, with non-human animals' interests ignored in the service of human wants.

10. Visitors [of the Bamboo Forest Coffee Shop] can also find a mongoose in a cage. The mongoose is just a kind of 'display' for tourists who are curious about the mongoose's appearance. (KP064, 2015)
11. Tourists [of agro-tourism in Karanganyar Village, East Java] can continue to a kind of resting place where they can sip luwak coffee surrounded by a mongoose breeding farm. (KP040, 2013)
12. There [Alas Harum Agro-tourism, Bali], visitors can taste freshly processed luwak coffee, even see the civets themselves and learn about the process of making the coffee. (KP156, 2022)

Moorhouse et al. (2015) studied wildlife tourism and found that the majority of such tourism negatively impacted wildlife and/or the environment generally. These researchers included displays of caged civets among the negative instances of wildlife tourism. Along the same lines, Hooper (2022) reported that tourists in Bali frequently complained about cages unsuitable for supplying basic necessities to civets, adding that most of the tourists who visited coffee farms had a positive experience with scenery and coffee but disliked caged civets, leading to negative reviews and boycott calls due to concerns about civet exploitation for photo opportunities.

A significant ethical dilemma confronts producers, promoters and consumers of civet coffee. Despite the industry's success in attracting tourists, it encounters opposition because of tourists' worries about civet welfare. Carder et al. (2016) rated the welfare of caged civets in sixteen locations on eight factors – mobility, hygiene, surfaces, shelter, noise, food, water and social interactions – and found conditions uniformly inadequate. Furthermore, Hooper (2022) argued that civet involvement in agro-tourism lacks conservation justification and does not even claim to serve that purpose.

In these excerpts, a prevalent ideology is observed to embrace an ambivalent discourse. These discourses aim to address ecological issues stemming from destructive discourses, yet they often originate from the same societal framework and can be influenced by political and commercial interests (Stibbe, 2021).

Milstein (2009) provided an example of ambivalent discourse in the context of zoos. Throughout history to the present, zoos have consistently revolved around two fundamental themes: 'gaze' and 'power.' Despite zoos' attempts to advocate for endangered species and ecosystems, they frequently perpetuate dominant discourses associated with control, 'othering' (different from or inferior to human animals) and exploitation.

In a similar way, the excerpts highlight civets being used as tourist attractions on coffee plantations. These animals are confined in cages for public viewing, treating them as sources of human entertainment and reflecting an anthropocentric perspective that prioritizes human desires over animal welfare.

Production process

This section of the chapter contains and comments on excerpts from the two newspapers, providing glimpses into the representation of civets and their role in the production of civet coffee.

13. The coffee fruit that the luwak swallows will undergo a fermentation process while it is in the animal's stomach. After being digested by the civet, the coffee fruit turns into beans which are excreted with the animal's feces. (KP002, 2009)
14. The popularity and huge profits of luwak coffee have led to the mushrooming of civet husbandry in areas across Bali. Luwak coffee has become a prime commodity and its production has developed into one of the island's attractive tourist destinations, with many tourists eager to see more closely the production method of the most expensive coffee in the world. (JP012, 2016)
15. Some people believe that the digestion and fermentation process caused by stomach acid in the civet is what enriches the taste of civet coffee. Some critics agree that this is just a gimmick that produces very bad espresso. Unfortunately, many farms breed mongooses specifically. They are kept in cages and force-fed to produce civet coffee. The price of civet coffee has become very expensive due to the fermentation process of the coffee beans in the civet's stomach. (KP127, 2020)

Croney and Reynnells (2008) asserted that language about the farm animal industry often conceals the negative impacts of animal production on the other-than-human world and the environment through deliberate use of grammar and vocabulary, particularly when discussing animal slaughter and other aspects of the production processes for animal-based foods, as transparency regarding these processes could raise public concern, leading to consumer aversion and subsequent economic losses.

Glenn (2004) explained that in discourses disseminated by the producers of animal-based foods, non-human animals are portrayed as inanimate objects and commodities, valued solely as products and producers for human use, offering an anthropocentric ideology. In the case of civets, it can be seen that even though they are mentioned as essential actors in the production of civet coffee, civets are not subjects in control of their destinies but rather objects controlled by humans (Glenn, 2004; Satyaningtijas et al., 2015).

Civet coffee: Taste, quality and price

This section of the chapter contains excerpts from the two newspapers that show how civet coffee is represented in terms of taste, quality and price.

16. Then there is luwak coffee which is said to be the most delicious and most expensive coffee in the world. (KP029, 2011)
17. [. . .] luwak coffee has a perfect taste and special aroma that is widely acclaimed by coffee lovers across the globe. The natural selection process of the coffee beans, which involves a civet's skill to select quality berries, is the main factor behind the perfect flavor of luwak coffee. (JP012, 2016)
18. When harvested in the wild [civet coffee], 500 grams of coffee beans can reach US$600, equivalent to IDR 9.2 million. (KP177, 2023)

In the market for meat from non-human animals, such as chickens and fishes, consumers usually focus on price and on experience characteristics such as taste, texture and appearance (Davidson et al., 2003; McEachern & Schröder, 2002). However, scarcity and high price can sometimes be an attraction (Shi et al., 2020). In the case of civet coffee, the focus is on its distinct taste and quality rather than affordability. Its high price and unique production make it an exclusive and luxurious product for consumers, a situation that can be compared with other industries, for example, the fashion industry.

In an analysis conducted regarding customer perceptions of civet coffee, Bayunitri (2017) suggested that customers believe that the elevated cost of civet coffee reflects its superior quality, attributed to the unique and intricate production process, unlike that of regular coffee. The civet's role in coffee production is commercialized and associated with concepts of quality and exclusivity within the market. Thus, nature is harnessed for commercial gain.

Achabou (2020) studied consumers' views on animal welfare issues concerning luxury products and found that some consumers preferred animal-based materials from so-called ethical farms, while others were more inclined to purchase synthetic materials. However, Achabou noted that the use of synthetics can lower the perceived value of luxury items. In the context of civet coffee, several studies have reported attempts to artificially ferment coffee and avoid the use of civets (Fauzi & Hidayati, 2016; Hadipernata & Nugraha, 2012; Raveendran & Murthy, 2021; Sudarti et al., 2020; Usman et al., 2015). However, the authors of the current chapter wonder whether consumers would favour this synthetic option over the traditional non-human animal-produced one.

Stibbe (2021) proposed the prevalence of consumerist ideology in which many consumers aim to maximize their own satisfaction through their purchases. They emphasize extrinsic values, that is, where value is placed on what people *have* (e.g. money, homes, consumer goods and country club memberships) rather than intrinsic values, that is, what people *are* (e.g. the skills and knowledge they possess and how they use these to benefit others). This consumerist ideology promotes increased use of natural resources in order to satisfy expanding consumer demand and opposes efforts to decrease consumption (Stibbe, 2021). Consumerist ideology is evident in the above excerpts, as the emphasis lies with the products humans can consume, not with the non-humans involved in the production of these products.

Civets as machines and workers

Metaphors are another type of story proposed by Stibbe (2021). The following excerpts use metaphors to characterize civets as machines and workers.

19. For good coffee quality, the health of the civet as a 'grinding machine' must be considered. The nutritional content of the food given must be considered. (KP027, 2011)
20. Their [civets] utility hadn't been explored at all as coffee 'fermenters' and their fecal matter was a mere inconvenience to the villager, as is dog crap to the jogger in New York City.' (JP005, 2015)
21. Pujo [a coffee farmer from Kendal, Central Java] said that he also produces luwak coffee. There are around 91 pet mongooses, which are used as producers. (KP094, 2017)

Metaphor and framing share similarities in their ability to shape perceptions and understanding. In the context of metaphor, the 'source frame' serves as the foundation for illustrating a more abstract concept by drawing from a distinct and often familiar sector of life (Stibbe, 2021). For example, the metaphor 'Life is like a ship sailing in a storm' employs a source frame of a ship navigating turbulent waters to represent life's challenges and difficulties. This source frame, rooted in concrete and tangible imagery, aids in making abstract ideas more relatable. Moreover, it's important to note that these metaphors extend their influence beyond the realm of language and can significantly impact how people perceive and interact with various subjects, including their relationships with non-human animals.

In the above excerpts, civets are portrayed metaphorically as workers or even machines. Hooper et al. (2022) emphasized that civets are turned into commercial objects, and the entire process mechanizes their bodies along a production line, starting from the mouth and ending at the anus. Stibbe (2021) explored the 'Nature is a machine' metaphor, commonly viewed as harmful to nature because it justifies the exploitation of nature, portraying this exploitation as normal, just as no one would question the use of machines. Such metaphors justify, normalize and propagate the mistreatment and undervaluation of nature, including civets.

Farming of non-human animals

The two excerpts in this section of the chapter are notable for the explicit connection between the treatment of civets and the treatment of other non-human animals that are used in the production of food and drink for humans.

22. Bayu [Bayu Krisnamurthi, Indonesia Deputy Minister of Trade] indicated that the accusations levelled at Indonesia were just a trade war, as the market for Indonesian luwak coffee was increasingly growing. He further said that the public must understand the different processes of luwak coffee production. '(whether) There is a violation of animal [civet] welfare rights or not, it should not be generalized that it [the production] is bad. This [production] is the same as a dairy cow being fed and milked,' said Bayu at the Ministry of Trade office, Jakarta, Friday (15/11/2013). (KP046, 2013)

23. [Note: in 2013, it was reported that Harrods, a high-end department store in London, had stopped selling civet coffee following allegations of mistreatment of civets (Anne, 2013).] Eggs, meat, fish and all dairy products are basically produced through animal cruelty. We put livestock in cages, raise them in very bad conditions and then kill them to bring food to our table. Common sense would seem to dictate that luwak coffee should not be excluded by Harrods if other animal products continue to be sold as well. Will Harrods take the same action for other similar products? [. . .] It is clear that producing food by killing or abusing animals is a stain on human civilization. But there is little that can be done about the issue due to the food commodification regime, which was pioneered by the western world. (JP002, 2013)

Previously, scholars such as Dunayer (2001) and Kunst and Hohle (2016) have noted a tendency to label products from non-human animals with names that conceal the products' animal origins in order to dissociate the product from the animals of origin. For example, 'Living animals are called cows, pigs, and calves, but once dead and prepared for food they become beef, pork and veal' (Kunst & Hohle, 2016, p. 759). The celebrity musician and vegan Paul McCartney is often quoted as asserting that 'If slaughterhouses had glass walls, everyone would be a vegetarian' (PETA2TV, 2013). Loughnan et al. (2010) coined the term 'meat paradox', that is, people want to eat meat, but they also want to be kind to non-human animals.

To make it very difficult for the public to see what goes on inside the production of animal-based foods, in other words, to ease the meat paradox, some countries have what are pejoratively called ag-gag laws (Gelber & O'Sullivan, 2021), that is, laws that criminalize the making of videos and other media showing conditions at slaughterhouses and at the facilities, known as concentrated animal feeding operations, where non-human animals are born and prepared to play their involuntary role in the production of food for humans (Fox, 2023). Stibbe (2021) classified such efforts as 'erasure' (i.e. the removal of information that is inconvenient to the purposes of those creating a text). Bray et al. (2016) reported a survey of parents as to what they told children about the origins of the meat they were eating.

However, Kunst and Hohle discussed an exception to this erasure of the animals used for animal-based foods. This exception involves the case of dairy cows, who are sometimes featured in the marketing of dairy products. Images of cows are used as the dairy industry perpetuates the myth that dairy cows live a contented life, as in the children's song 'Old MacDonald Had a Farm', and that milking benefits the cows. In reality, however, dairy cows suffer such forms of mistreatment as forced impregnation, the immediate and permanent post-birth separation from their calves, and death once the cows' milk production declines (Puech, 2023). In this case, the dairy cows are not erased at all; rather, they are highlighted but in a specific way, what Grauerholz (2007, p. 344) termed 'cutification' (i.e. the use of illustrations of cute-looking animals to attract consumers and to banish the idea of cruelty from consumers' minds). Sometimes anthropomorphism, that is, giving human-like characteristics to non-humans, is used in this process (e.g. the ability to talk in human languages).

With civet coffee, the civets' situation may be more analogous to that of dairy cows or egg-laying hens compared to the situation of cows and chickens raised specifically for meat. However, whereas dairy cows and egg-laying hens face death due to their diminishing ability to produce milk and eggs respectively, civets' ability to defecate may be longer-lasting relative to their life spans. What happens to ageing caged civets may be an area for future research.

Conclusion

This chapter has applied insights from ecolinguistics to study the representation of civets in articles about civet coffee in two of Indonesia's leading newspapers. The analysis found many similarities between the representation of civets and that of other non-human animals used in the production of consumer goods. However, the role of civets in the production of civet coffee did suggest certain unique features of the civets' situation.

Limitations and suggestions for future research

The current study suffered from a number of limitations. These include the fact that only newspapers were studied, not other forms of media, and only words, not visuals, from those newspapers. Also, texts produced by the coffee companies and various people involved in that industry were not considered. Third, the study was confined to Indonesia, even though similar coffee products are produced and/or sold in other countries.

Future researchers might be able to address some of the limitations of the current research. For instance, researchers could look at the representations of civets and civet coffee in visual texts. It would also be useful to understand the perspectives of small farmers and agricultural workers involved with civet coffee. Additionally, the views of consumers in Indonesia and other countries with regard to civet coffee could be studied.

Call to action

Language plays key roles in the construction of society, including the economic systems of a society. Ecolinguistics empowers people to better understand the roles of language. Unfortunately, injustices towards fellow humans and other fellow animals exist within societies. As ecolinguistics develops, it offers ideas on how language can facilitate resistance to these injustices, and the ideologies and systems that breed and defend injustice. Furthermore, language resources are needed to build alternative, less destructive, economic systems (Stibbe, 2021). Thus, ecolinguists are challenged to not only promote understanding of the world but also to engage our fellow humans in creating a world of justice for all humans and, by necessity, for the more-than-human world as well.

References

Achabou, M. A. (2020). Is animal welfare a central issue for consumers of luxury goods? *Natural Resources Forum*, *45*(1), 18–36. https://doi.org/10.1111/1477-8947.12211

Agmarasi, S. (2016, December 16). *Berapa harga yang pantas untuk kopi luwak?* [What is a reasonable price for luwak coffee?]. Kompas.com. https://travel.kompas.com/read/2016/12/16/180500627/berapa.harga.yang.pantas.untuk.kopi.luwak

Agnew, R. (1998). The causes of animal abuse: A social psychological analysis. *Theoretical Criminology*, *2*(2), 177–210. https://doi.org/10.1177/1362480698002002003

Anne. (2013, October 18). *Harrods makes humane choice on civet coffee*. PETA. Retrieved from https://www.peta.org.uk/blog/harrods-makes-humane-choice-on-civet-coffee

Barrow, I. (2017). *The East India Company, 1600–1858: A short history with documents*. Hackett Publishing Company.

Bayunitri, B. I. (2017). Customer perceptions of product luwak coffee based on marketing mix elements perspective (Customer survey in Bandung). *Proceedings of the 2017 International Conference on Organizational Innovation*. Atlantis Press. https://doi.org/10.2991/icoi-17.2017.33

Bliss, S. (2017). Coffee interconnections. *Geography Bulletin, 49*(4), 55–71. https://search.informit.org/doi/10.3316/ielapa.129417376394920

Bray, H. J., Zambrano, S. C., Chur-Hansen, A., & Ankeny, R. A. (2016). Not appropriate dinner table conversation? Talking to children about meat production. *Appetite, 100*, 1–9. https://doi.org/10.1016/j.appet.2016.01.029

Cahill, C. (2019). [Un]becoming a resource: Translating the nature of civets in Indonesia. *Ethnos, 85*(1), 100–117. https://doi.org/10.1080/00141844.2018.1536072

Carder, G., Proctor, H., Schmidt-Burbach, J., & D'Cruze, N. (2016, May). The animal welfare implications of civet coffee tourism in Bali. *Animal Welfare, 25*(2), 199–205. https://doi.org/10.7120/09627286.25.2.199

Croney, C., & Reynnells, R. (2008). The ethics of semantics: Do we clarify or obfuscate reality to influence perceptions of farm animal production? *Poultry Science, 87*(2), 387–391. https://doi.org/10.3382/ps.2007-00404

D'Cruze, N., Toole, J., Mansell, K., & Schmidt-Burbach, J. (2014, March 13). What is the true cost of the world's most expensive coffee? *Oryx, 48*(2), 170–171. https://doi.org/10.1017/s0030605313001531

Davidson, A., Schröder, M. J., & Bower, J. A. (2003). The importance of origin as a quality attribute for beef: Results from a Scottish consumer survey. *International Journal of Consumer Studies, 27*(2), 91–98. https://doi.org/10.1046/j.1470-6431.2003.00288.x

Deshmukh, A. (2021, October 1). *The world's top coffee producing countries*. Visual Capitalist. Retrieved from https://www.visualcapitalist.com/worlds-top-coffee-producing-countries

Djumena, E. (2013, March 3). *Nikmatnya kopi luwak* [Enjoy wild Luwak coffee]. Kompas.com. Retrieved from https://lifestyle.kompas.com/read/2013/03/18/16032679/Nikmatnya.Kopi.Luwak.Liar?

Dunayer, J. (2001). *Animal equality: Language and liberation*. Ryce.

Fauzi, M., & Hidayati, N. W. (2016). *Perubahan karakteristikkimia kopi luwak robusta in vitro denganvariasi lama fermentasi dan dosis ragi* [Changes in the chemical characteristics of robusta civet coffee in vitro with long fermentation variations and yeast doses]. *Proceedings of the Seminar Nasional Hasil Penelitian dan Pengabdian Masyarakat 2016*, Indonesia, 80–84.

Fox, L. (2023). The intersectionality of environmental injustice, other societal harms, and farmed animal welfare. environmental justice. *Environmental Justice, 17*(2), 1–9. https://doi.org/10.1089/env.2021.0125

Fridayani, N. (2020, December 8). *5 kopi termahal di dunia, ada yang berasal dari sisa kotoran luwak dan gajah* [The 5 most expensive coffees in the world, some come from civet and elephant droppings]. Kompas.com. Retrieved from https://www.kompas.com/food/read/2020/12/08/154248475/5-kopi-termahal-di-dunia-ada-yang-berasal-dari-sisa-kotoran-luwak-dan-gajah

Gelber, K., & O'Sullivan, S. (2021). Cat got your tongue? Free speech, democracy and Australia's 'ag-gag' laws. *Australian Journal of Political Science, 56*(1), 19–34. https://doi.org/10.1080/10361146.2020.1799938

Glenn, C. B. (2004). Constructing consumables and consent: A critical analysis of factory farm industry discourse. *Journal of Communication Inquiry*, *28*(1), 63–81. https://doi.org/10.1177/0196859903258573

Grauerholz, L. (2007). Cute enough to eat: The transformation of animals into meat for human consumption in commercialized images. *Humanity & Society*, *31*(4), 334e354. http://dx.doi.org/10.1177/016059760703100404

Hadipernata, M., & Nugraha, S. (2012). Identifikasi fisik, kimia dan mikro biologi biji kopi luwak sebagai dasar acuan teknologi proses kopi luwak artificial [Physical, chemical and microbiological identification of luwak coffee beans as a basic reference for artificial luwak coffee process technology]. *Jurnal Kementer Pertan*, *372*, 117–121.

Hameed, A. (2021). An ecolinguistic perspective on framing of animals in Quranic discourse. *Arab World English Journal for Translation and Literary Studies*, *5*(3), 113–124. https://doi.org/10.24093/awejtls/vol5no3.9

Harris, J. B. C., Tingley, M. W., Hua, F., Yong, D. L., Adeney, J. M., Lee, T. M., Marthy, W., Prawiradilaga, D. M., Sekercioglu, C. H., Suyadi, Winarni, N., & Wilcove, D. S. (2017). Measuring the impact of the pet trade on Indonesian birds. *Conservation Biology*, *31*(2), 394–405. https://doi.org/10.1111/cobi.12729

Hooper, J. (2022). Cat-poo-chino and captive wildlife: Tourist perceptions of Balinese kopi luwak agrotourism. *Society & Animals*, *32*, 229–249. https://doi.org/10.1163/15685306-bja10094

Hooper, J., Linna, M., Kassinen, S., & Salvage, J. (2022). Technologies, bodies and faecal matters: Embodied empathy with coffee producing civets. *Transpositions*, *1*(2), 73–94. https://doi.org/10.14220/trns.2022.1.2.73

International Ecolinguistics Association. (n.d.). *About*. Retrieved from https://www.ecolinguistics-association.org

Iskandar, B. S., Iskandar, J., & Partasasmita, R. (2019). Hobby and business on trading birds: Case study in bird market of Sukahaji, Bandung, West Java and Splendid, Malang, East Java (Indonesia). *Biodiversitas*, *20*(5). https://doi.org/10.13057/biodiv/d200522

Ivory, M. (2021, March 23). *Tertarik memelihara musang? Perhatikan ini* [Interested in keeping weasels? Watch this]. Kompas.com. Retrieved from https://lifestyle.kompas.com/read/2021/03/23/184411820/tertarik-memelihara-musang-perhatikan-ini?

Jamal Said, N. (2022, September, 10). *Daftar 5 kopi termahal di dunia, salah satunya berasal dari Indonesia* [List of the 5 most expensive coffees in the world, one of which comes from Indonesia]. Kompas.com. Retrieved from https://money.kompas.com/read/2022/03/06/194435226/daftar-5-kopi-termahal-di-dunia-salah-satunya-berasal-dari-indonesia?page=all

Keil, J. (2016). Relationships, attitudes and emotions of humans towards nonhuman primates in Indonesia. In A. Höing & A. M. Bennett (Eds.), *Humans and animals: Intersecting lives and worlds* (pp. 55–63). Brill.

Koknaroglu, H., & Akunal, T. (2013). Animal welfare: An animal science approach. *Meat Science, 95*(4), 821–827. https://doi.org/10.1016/j.meatsci.2013.04.030

Kunst, J. R., & Hohle, S. M. (2016). Meat eaters by dissociation: How we present, prepare and talk about meat increases willingness to eat meat by reducing empathy and disgust. *Appetite, 105*, 758–774. https://doi.org/10.1016/j.appet.2016.07.009

Lachenmeier, D. W., & Schwarz, S. (2021). Digested civet coffee beans (Kopi Luwak)—An unfortunate trend in specialty coffee caused by mislabeling of Coffea liberica? *Foods, 10*(6), 1329. https://doi.org/10.3390/foods10061329

Langit Nariswari, S. (2023, July 10). *10 makanan termahal di dunia, termasuk kopi luwak Indonesia* [10 most expensive foods in the world, including Indonesian luwak coffee]. Kompas.com. Retrieved from https://lifestyle.kompas.com/read/2023/07/10/101452820/10-makanan-termahal-di-dunia-termasuk-kopi-luwak-indonesia?page=all

Loughnan, S., Haslam, N., & Bastian, B. (2010). The role of meat consumption in the denial of moral status and mind to meat animals. *Appetite, 55*(1), 156–159. http://dx.doi.org/10.1016/j.appet.2010.05.043

Lynn, G., & Rogers, C. (2013, September 13). *Civet cat coffee's animal cruelty secrets.* BBC. Retrieved from https://www.bbc.com/news/uk-england-london-24034029

Marocknroll (n.d). *Goats on the trees trip from Agadir.* Retrieved from https://marocknrolltours.com/product/goats-on-the-trees-trip-from-agadir/

McEachern, M. G., & Schröder, M. J. A. (2002). The role of livestock production ethics in consumer values towards meat. *Journal of Agricultural & Environmental Ethics, 15*(2), 221–237. https://doi.org/10.1023/a:1015052816477

Milstein, T. (2009). "Somethin' tells me it's all happening at the zoo": Discourse, power, and conservationism. *Environmental Communication, 3*(1), 25–48. https://doi.org/10.1080/17524030802674174

Ministry of Agriculture of the Republic of Indonesia. (2015). Peraturan Menteri Pertanian Nomor 37 Tahun 2015 tentang Cara Produksi Kopi Luwak Melalui Pemeliharaan Luwak Yang Memenuhi Prinsip Kesejahteraan Hewan [Minister of agriculture regulation number 37 year 2015 concerning methods for production of Luwak Coffee through Civet rearing that meets animal welfare principles]. Retrieved from https://jdih.pertanian.go.id/fp/peraturan/detail/716

Moorhouse, T. P., Dahlsjö, C. A. L., Baker, S. E., D'Cruze, N. C., & Macdonald, D. W. (2015). The customer isn't always right—Conservation and animal welfare implications of the increasing demand for wildlife tourism. *PLOS ONE, 10*(10), e0138939. https://doi.org/10.1371/journal.pone.0138939

Muzaifa, M., Hasni, D., Rahmi, F., & Syarifudin. (2019). What is kopi luwak? A literature review on production, quality and problems. *IOP Conference Series: Earth and Environmental Science, 365*(1), 012041. https://doi.org/10.1088/1755-1315/365/1/012041

National Parks Board. (2022, March 7). *Civets.* National Parks Board of Singapore. Retrieved from https://www.nparks.gov.sg/gardens-parks-and-nature/dos-and-donts/animal-advisories/civets

Nijman, V., & Nekaris, K. A. I. (2017). The Harry Potter effect: The rise in trade of owls as pets in Java and Bali, Indonesia. *Global Ecology and Conservation, 11*, 84–94. https://doi.org/10.1016/j.gecco.2017.04.004

Nursastri, S. A. (2017, May 28). Melancong ke Candi Pawon, jangan lupa minum kopi luwak [Traveling to Pawon Temple, don't forget to drink kopi luwak]. Kompas.com . Retrieved from https://travel.kompas.com/read/2017/05/28/120500427/melancong.ke.candi.pawon.jangan.lupa.minum.kopi.luwak

Parry, J. (2004). WHO queries culling of civet cats. *BMJ, 328*(7432), 128. https://doi.org/10.1136/bmj.328.7432.128-b

Peresthu, A. (2013, October 31). *Luwak coffee: From animal welfare to national heritage.* The Jakarta Post. Retrieved from https://www.thejakartapost.com/news/2013/10/31/luwak-coffee-from-animal-welfare-national-heritage.html

PETA. (2022). *Civets suffering for cruel coffee. People for the Ethical Treatment of Animals.* PETA. Retrieved from https://www.peta.org/features/civets-suffering-cruel-coffee.

PETA2TV. (2013, May 15). *Paul McCartney narrates PETA2's "Glass Walls"* [Video]. YouTube. Retrieved from https://www.youtube.com/watch?v=a5iwL6zQ8bk

Prajna, D., Álvarez, M., Barea-Sepúlveda, M., Calle, J. L. P., Suhandy, D., Setyaningsih, W., & Palma, M. (2023). Enhanced differentiation of wild and feeding civet coffee using near-infrared spectroscopy with various sample pretreatments and chemometric approaches. *Horticulturae, 9*(7), 778. https://doi.org/10.3390/horticulturae9070778

Puech, L. T. (2023). Big dairy: Global tactics and political influence. *Animal Ethics Review, 3*(1), 51–59. https://doi.org/10.31009/aer.2023.v3.i1.04

Rachmawati, I. (2016, August 11). *Kopi luwak di Banyuwangi dihargai Rp 2 juta per kilogram* [Luwak coffee in Banyuwangi is priced at IDR 2 million per kilogram]. Kompas.com. Retrieved from https://regional.kompas.com/read/2016/08/11/15301161/kopi.luwak.di.banyuwangi.dihargai.rp.2.juta.per.kilogram.?page=all

Raveendran, A., & Murthy, P. S. (2021). New trends in specialty coffees – "The digested coffees." *Critical Reviews in Food Science and Nutrition, 62*(17), 4622–4628. https://doi.org/10.1080/10408398.2021.1877111

Regan, T. (2003). *Animal rights, human wrongs: An introduction to moral philosophy.* Rowman & Littlefield Publishers.

Roos, R. (2004). *WHO sees more evidence of civet role in SARS.* CIDRAP (Center for Infectious Disease Research & Policy) News: https://www.cidrap.umn.edu/sars/who-sees-more-evidence-civet-role-sars

Satyaningtijas, A., Kusumorini, N., Fachrudin, M., &., P. (2015). Profil Leukosit, diferensialleukosit, dan indeksstres luwak jawa (paradox rushermaphroditus) (leucocyte count, leucocyte differentiation, and stress index of common palm civets. (Paradoxurus Hermaphroditus). *JurnalVeteriner, 15*(4), 487–493. https://Ojs.Unud.Ac.Id/Index.Php/Jvet/Article/View/13227

Shi, X., Li, F., & Chumnumpan, P. (2020). The use of product scarcity in marketing. *European Journal of Marketing, 54*(2), 380–418. https://doi.org/10.1108/EJM-04-2018-0285

Singer, P. (2015). *Animal liberation*. Bodley Head Childrens.

Sollund, R. (2011). Expressions of speciesism: The effects of keeping companion animals on animal abuse, animal trafficking and species decline. *Crime, Law and Social Change, 55*(5), 437–451. https://doi.org/10.1007/s10611-011-9296-3

Stibbe, A. (2021). *Ecolinguistics: Language, ecology and the stories we live by* (2nd ed.). Routledge.

Sudarti, B., S., Prastowo, S. H. B., Prihandono, T., Maryani, & Handayani, R. D. (2020). Optimizing lactobacillus growth in the fermentation process of artificial civet coffee using extremely- low frequency (ELF) magnetic field. *Journal of Physics: Conference Series, 1465*(1), 012010. https://doi.org/10.1088/1742-6596/1465/1/012010

Sullivan, M. (2014, August 20). No. 1 most expensive coffee comes from elephant's no. 2. *National Public Radio*. Retrieved from https://www.npr.org/sections/thesalt/2014/08/20/340154271/no-1-most-expensive-coffee-comes-from-elephants-no-2

Usman, D., Suprihadi, A., & Kusdiyantini, E. (2015). Fermentasi kopi robusta (coffe acanephora) menggunakanisolat bakteria samlaktatdari feces luwak denganperlakuan lama waktuinkubasi. *Jurnal Akademika Biologi, 4*(3), 31–40. https://ejournal3.undip.ac.id/index.php/biologi/article/download/19417/18416

Wakhid, S., Sarno, R., Sabilla, S. I., & Maghfira, D. B. (2020). Detection and classification of Indonesian civet and non-civet coffee based on statistical analysis comparison using E-Nose. *International Journal of Intelligent Engineering & Systems, 13*(4), 56–65.

World Animal Protection. (n.d.). *Fact File: Asian palm civet*. Retrieved from https://www.worldanimalprotection.org/sites/default/files/media/int_files/cive-factfile.pdf

World Coffee Research. (2023, October 18). *Ensuring the future of coffee*. https://worldcoffeeresearch.org/

Zhdanava, A., Kaur, S., & Rajandran, K. (2021). Representing nonhuman animals as equals: An ecolinguistic analysis of vegan campaigns. *Journal of World Languages, 7*(1), 26–57. https://doi.org/10.1515/jwl-2021-0003

Index

www.ingramcontent.com/pod-product-compliance
Ingram Content Group UK Ltd.
Pitfield, Milton Keynes, MK11 3LW, UK
UKHW020805290926
15470UKWH00003B/955

9 781350 523807